I0062457

THE "COMMON SENSE" OF CRYPTOCURRENCY

by Robin Bloor

Little Crow
Press

**Little Crow
Press**

Austin, Texas

THE "COMMON SENSE" OF CRYPTOCURRENCY

Copyright © 2019 by The Little Crow Press

All Rights Reserved.

Without limiting the rights under copyright reserved above, no part of this publication may be reproduced, stored in or introduced into a retrieval system, or transmitted, in any form or by any means (electronic, mechanical, photocopying, recording or otherwise), without the prior written permission of both the copyright owner and the above publisher of the book. Please do not participate in or encourage piracy of copyrighted materials in violation of the author's rights. Purchase only authorized editions.

The author has made every effort to provide accurate Internet addresses in this work at the time of publication. Neither the publisher nor author assumes any responsibility for errors or changes that occur after publication. Further, the publisher and author have no control over and do not assume responsibility for third-party websites or their content.

First Edition: May 2019

ISBN: 978-0-9966299-1-1

Printed in the United States of America

I offer nothing more than simple facts, plain argument and common sense

~ Thomas Paine

THE
"COMMON SENSE"
OF
CRYPTOCURRENCY

CONTENTS

Chapter 1

The Ghost of Tom Paine

"An army of principles can penetrate where an army of soldiers cannot."

~ Tom Paine

A technology wave, insane investment profits and wacky business ideas. It seemed like the second coming of the dot-com era. The parallels were all there. You could guess what would happen next.

- Seasoned commentators dismissed the cryptocurrency phenomenon as a fad, a fraud, a Ponzi scheme. They did many times, years before cryptocurrency values shot into the stratosphere.

- The vast majority of the upstart start-ups soon began to fade away like morning dew. Over a thousand cryptocurrencies evaporated and hundreds more are now on life support.

- All the wacky business ideas would fail. Of course, they would. Wacky business ideas usually do.

- Some of these new commercial ventures would prove durable. No surprise there. Innovative technology, clever ideas, and good planning sometimes breed success. Nevertheless, as I write these words, there are no real superstars among the crypto businesses.

- Cryptocurrency technology—the blockchain—will bring change. Exactly how may be unclear, but change it will surely bring.

Many of us who soldiered through the dot-com era see the cryptocurrency landscape as an exercise in déjà vu. It's the dot-com revolution repeating, endowed with different duds. Nevertheless, there is something distinctly different about cryptocurrency, which indicates a distinctly different outcome.

Dial yourself back to the dawn of the Internet. You sat at your PC one day and stumbled into a new app on your screen, called a *browser*. How exciting! And it was exciting. You could "explore" a vast new digital

1

landscape, populated by Yahoo!, Amazon.com, BlueMountainArts.com, About.com, Weather.com, AskJeeves.com and more. You set out across this uncharted territory armed with nothing more than the AltaVista Search engine, or one of its cousins.

You learned where to go for the news, how to buy from websites, how to use web-based email. And then anything new emerged—Internet banking, stock trading, social networks, Dropbox—you added it in and thought little of it. You were a sponge and you soaked up whatever the nascent Internet provided. The dot-com era changed your life to be sure.

But the cryptocurrency revolution is different. While you may have joined the investment mania, aside from learning how to use a crypto wallet, your day-to-day existence has seen no great changes. Currently, the compelling cryptocurrency applications are immature. That's one side of the coin. The other side is this: the cryptocurrency revolution is a true revolution—a basic human *right* is at stake.

Like all human rights, we will not win it without a fight. And our struggle will be ineffective until we—men and women of the 21st century--recognize and acknowledge this new right as immutable and fundamental. It is this:

We own our personal data.

Enter Tom Paine

We will fight for possession of our data. Looking through that lens, the impending revolution is not defined by cryptocurrency and the block-chain; it's wider than that. It is a Data Revolution. Cryptocurrency is part and parcel of it, in fact crucial to it, and it will play a major role. Ultimately, this revolution is about our data. When the dust settles, we must, we will have inalienable rights over our personal data.

Already there have been skirmishes. Our situation has parallels with another, older insurgence. If you're a history buff, consider the American Revolution after The Boston Massacre of 1770, after The Boston Tea Party of 1773, and even after the "The shot heard round the world," when the Minutemen and redcoats clashed at Lexington and Concord. Think about the month of January 1776 when Tom Paine published *Common Sense* and demonstrated that his pen was as mighty as the musket.

He wrote:

"Europe, and not England, is the parent country of America. This new world hath been the asylum for the persecuted lovers of civil and religious liberty from every part of Europe. Hither they have fled, not from the tender embraces of the mother, but from the cruelty of the monster; and it is so far true of England, that the same tyranny which drove the first emigrants from home, pursues their descendants still."

Common Sense, a 48-page pamphlet, quickly became a best-seller. The pamphlet sold an estimated 100,000 copies to a population of 2.5 million. It is still in print and is said to be the best selling American book of all time. No surprise really, it inspired the American Revolution.

At the time of the American Revolution most of the population of the Thirteen Colonies thought of themselves as British subjects, plagued by a mother country determined to impose unpopular laws and taxation on them. From our perspective the transformation of the Colonies into a new republic might seem the natural course of action, the obvious outcome requiring merely a well-led revolt against a foolish army of redcoats.

But in 1770, the creation of a republic was not so easy to envisage. Britain was a superpower, revolt was mutiny, and monarchies had never been successfully replaced by republics. Oliver Cromwell's attempt to establish a republic in England in 1653 fizzled out with Cromwell's death. The English could think of nothing better to do than restore the monarchy. The colonists had every reason to expect their revolution would sputter its way to an equally pathetic conclusion.

Common Sense is both logical and impassioned. Paine pointed out the idiocy of people being subject to laws they never created; he identified flaws in the British system of government, particularly as it affected the colonists; and he argued strongly against heredity monarchy—against this imbalance of power. His work was a tour de force and changed people's minds, and in doing so, it changed history.

Right now, we face a similar imbalance of power to the one Tom Paine articulated so effectively. The imbalance at issue for us, however, is not in the government of people, it is in the government of data. When you examine it closely, you realize that the governance of data has implications for one's personal life, one's social life…in truth, for one's life in all its

3

dimensions—social, business, health, financial, legal. It touches everything and will ultimately touch the government of people.

It is probably not clear to you that this is so. But it will become clearer as we discuss the nature of data, the nature of money, the nature of commerce and the nature of government.

A Data Retrospective and Prospective

Recorded history began around the 4th millennium BC, written in Sumerian cuneiform script. There is not an extensive amount of it. What there is provides evidence of bookkeeping, recipes, medical remedies, scripture and historical records. None of these things are surprising. Whether there was much commerce prior to that time is difficult to demonstrate. There is evidence of tally sticks—sticks with count marks on them—which may go as far back as 30,000 years. But there is nothing to indicate how they were used.

The importance of recorded data is obvious. As the Chinese proverb says, "the palest ink is better than the best memory." However, the difficulties caused by the lack of technology to record data are not so obvious. We have forgotten. In the age of Gutenberg, which began over 500 years ago, print technology remade the world as the cost of reproducing data stored on paper began to collapse. And we have forgotten many of the difficulties caused by the limitations of that technology.

We now live in an age when computer technology in all its forms has become bewilderingly cheap. The technology was fully "data capable" years ago in the sense of being able to store and manipulate data of any kind. Nevertheless, it was not until the invention of blockchain technology that it became possible for people to own their data. It is now technically possible, as we shall explain later in this book.

However, "technically possible" is not good enough. People must understand the consequences of not owning their own data. And they must step forward and decide to take ownership. At the moment they have no inclination to do so, just as the colonists of the 13 states once had no inclination to revolt against the British crown.

But then the British crown became disturbingly oppressive and many of those colonists read a pamphlet that caused them to ponder the thorny question of freedom.

A Data Rights Revolution

According to the United States Historical Census, in 1775 the population of the Thirteen Colonies was 48.7% English and 14.4% Scottish or Scots-Irish, who would at the time have identified as British. The second largest ethnic group, accounting for 20% were African slaves and the remaining 16.9% were primarily European. The majority of the Founding Fathers were of English extraction, including the major players: Benjamin Franklin, George Washington, John Adams, James Madison, Alexander Hamilton and Thomas Jefferson—creatures of The Enlightenment, intent on creating a government of a different kind.

The American Revolution was unique—a band of idealists, born of the British Empire, opposing the hegemony of the English crown. Such a war was only feasible in a colonial territory with sufficient population and local resources to sustain its own army. The colonists had to want to battle with their English overlords. And they also had to carve out a new national identity—to think of themselves as American, not British.

Our position is similar. We are not Luddites, who wish to destroy the machinery of automation, any more than the colonists wished to destroy the English monarchy. We are the opposite.

We are early adopters of a new and powerful field of automation. We have colonized the territory, and we will use it to establish an entirely different sovereignty for our data. When enough people understand what is at stake, we will constitute an army, and we shall overturn the hegemony of the data landlords.

Data Wrongs

There have been big data robberies since Mark Zuckerberg was a boy (which is not so long ago). In recent years, they have been record-setting, courtesy of your friendly neighborhood Yahoo!. It announced two data breaches in 2016. The first, which hit the headlines in September 2016, was executed in late 2014, affecting 500 million Yahoo! user accounts. The second took place earlier, in late 2013, but remained unannounced until December 2016. The impressive score for that hack was 1 billion user accounts (rounded to the nearest billion). However, someone called for a recount, and in October 2017, when the results were announced, the final score was an impressive 3 billion. That was the whole population of Yahoo! users—roughly 75% of Internet users at the time.

Unimpressed by Yahoo!'s sloppy cyber security, credit bureau Equifax strove to go one better, and pulled it off when it was hacked for the data of 147 million U.S. residents. What Equifax failed to achieve in quantity—3 billion is a tough score to beat—it compensated for in quality.

The data stolen was far more valuable to hackers than Yahoo! data. It included a person's name, birthdate, Social Security number and, in some cases, address and driver's license number. That may not sound like much, but according to the pundits, it's exactly what a cyber thief needs to steal in order to hijack your precious identity and ruin your peace of mind for weeks, months, and years as you deal with the damage. And Equifax harvested your data—that they failed so abysmally to care for—without so much as a "by your leave."

Perhaps you imagine that the ruinous data breach destroyed Equifax. Not a bit of it. Certainly, the stock tanked—losing a third of its value on the announcement—but it recovered half of what it lost in the months that followed. Equifax actually *profited* from the hack. Advice to consumers in the wake of the hack was to "freeze your credit file to protect against abuse of your data"—and one in five U.S. consumers did exactly that. In doing so, they paid a freezing fee to the credit agencies (average cost $23). Wakefield Research estimated that this generated up to $1.4 billion in revenue for the big three credit agencies: Equifax, Experian and TransUnion.

And as for the US government, it has done nothing so far to punish Equifax, except to indict a few Equifax employees for insider trading when evidence emerged that they knew of the hack, and yet sold Equifax stock before the hack was announced.

Zuckerberg's Follies

Failing to protect personal data is one thing, spraying personal data around is quite another. In terms of data abuse, Facebook surpassed Yahoo and Equifax in their competitive race to the bottom. Reacting to those other record breaches, the down-trodden consumer probably just shrugged, and muttered philosophically "hacks have a habit of happening." But the Facebook data breach was a breach too far, because—if the reports are accurate—it wasn't a breach at all. Let us summarize:

- Cambridge Analytica wanted personal data in order to help the political campaign of Donald Trump. No surprises there, working on behalf of politicians and political parties was its business model.

- It engaged a firm called Global Science Research to build a downloadable Facebook app that paid Facebook users small rewards to take a "personality quiz" supposedly "for academic research."
- This app harvested the data of 270,000 users directly and, using Facebook friend links, eventually reaped the personal data of 87 million Facebook users, which Cambridge Analytica then analyzed and exploited.
- There is dispute as to whether this violated Facebook's terms (it probably did) but, reportedly, Facebook never policed those kinds of apps, so poaching data was not difficult.

When the news of this breach-or-non-breach broke, it pushed the outrage button of a larger number of people than either the Yahoo! or Equifax breaches had. This, it seems, had more to do with how the personal data was exploited (politically) than that it was stolen.

And just as happened with Equifax, when the Cambridge Analytica news broke in March 2018, Facebook's stock price fell on the floor, losing 17% of its value in a few days. But it dusted itself off and climbed right back up again and was later seen walking on air 33% higher than its low point and higher than it had ever been. It didn't matter that a "Delete Facebook" campaign took flight on Twitter, or that Mark Zuckerberg had to appear before Congress and meet with EU regulators. The Facebook stock price cares nothing for that; it cares only about ad revenues and user population growth, and by April 2018 neither of those figures were falling.

The Data Economy As Is

There are many businesses that do not want us to own and command our personal data. Facebook just happens to be the one that was caught red-handed and was publicly bathed in negative publicity. If you look at a list of the five largest companies, by market value, listed on the NYSE (on July 6th 2018), you can see that Facebook is merely the fifth.

Rank	Company	Value ($Bn)
1	Apple	923.9
2	Amazon	830.0
3	Alphabet	796.0
4	Microsoft	777.2
5	Facebook	588.1

	1970	1980	1990	2000	2010	2018
1	IBM	IBM	Exxon	Microsoft	Exxon Mobil	Apple
2	AT&T	AT&T	General Electric	General Electric	Microsoft	Amazon
3	General Motors	Exxon	IBM	Cisco	Walmart	Alphabet
4	Eastman Kodak	General Motors	AT&T	Walmart	Apple	Microsoft
5	Exxon	Amoco	Philip Morris	Exxon Mobil	Johnson & Johnson	Facebook
6	Sears Roebuck	Mobil	Merck	Intel	Proctor & Gamble	Alibaba
7	Texaco	General Electric	Bristol Myers	Lucent	IBM	Tencent Holdings
8	Xerox	Chevron	Dupont	IBM	JPMorgan Chase	Berkshire Hathaway
9	General Electric	Atlantic Richfield	Amoco	Citigroup	AT&T	JPMorgan Chase
10	Gulf Oil	Shell Oil	Bellsouth	AOL	General Electric	Exxon Mobil
						Source; Research Affiliates

Table 1. The USA's Most Valuable Companies, Decade By Decade

No doubt you have noticed that all five of those top companies earn their living from the Internet and mobile technology. That's where the gold is buried these days. The current dominance of that business sector is amplified when you notice that 6th and 7th on this list are Alibaba (often thought of as the Chinese Amazon) and Tencent Holdings (Chinese social networking, music, e-commerce, internet services, payment systems, smartphones, and multiplayer online games).

Take a look at *Table 1* and then think back to past decades. The trend is striking. In 1970, the top five business giants—GM, Exxon, Ford, GE, IBM—were spread out across multiple sectors. By 1980, oil companies were pre-eminent, due partly to the oil crises of the 1970s which drove up the price of oil. In 1990, we witnessed the rise of the pharmaceutical businesses. By 2000, it was all about the Internet and associated technologies. By 2010, a mixture of sectors seems to have re-established itself: Walmart, Exxon, Chevron, GE, Bank of America.

Then here we are, a mere 8 years later, and it looks as though a cruel culling took place that left only the Internet technology companies standing.

Look back over half a century and you can't find any year like 2018, when the top seven companies came from the same sector. The closest comparison is in 1980, when six of the top 10 companies were oil companies, and even then, IBM and AT&T still ruled the roost.

How Data Giants Got Their Chops

In 2006, Clive Humbly, a UK Mathematician working for the large UK supermarket Tesco was the first to claim, "Data is the new oil." His proclamation has since become a regularly repeated mantra, and it's true, in the sense that collections of data can be very valuable. But how come all this valuable data mining is only happening now?

We can see it as being provoked by a confluence of the following technology trends:

- The cost of data storage came crashing to the ground like a meteorite, year after year after year. It fell from about $1 million per megabyte in 1966 to $0.02 per gigabyte in 2018, and the costs are still falling.

- The costs of processing data, the cost of CPUs and GPUs also fell year after year after year, and did so even faster than a meteorite, with the cost per Gigaflop falling from $18.7 billion in 1961 to $0.03 in 2017.

- The advent of cloud computing in 2006 cut the costs further, making it possible to run computers less expensively by building huge data centers in areas of cheap electricity—classic economies of scale, applied directly to the cost of computing.

- The advent of massively parallel computing using Hadoop, Spark and other software technologies, coupled with cloud computing and rapidly falling computer costs, made it possible to statistically analyze extremely large collections of data in ways that had not previously been possible—and cheaper and faster.

It doesn't matter whether you fully understand the technology trends described above. When you boil it all down, it adds up to this: If you collect large amounts of data as part of your business, then the odds are that you'll be able to squeeze mountains of dollars out of it.

The Artificial Intelligence (AI) Dynamic

Because of the escalation of computer power, the capability of AI increases every year. In recent years, Stephen Hawking, Elon Musk and others expressed concern that AI could lead to human extinction—a kind of digital Frankenstein nightmare. The foreboding proposition is that AI technology will become obsessed with the task of self-improvement and improve itself far beyond human capability, and then, perhaps, swat us like flies.

In reality, better-than-human cognitive computer technology is already here; it's just not conveniently organized. In 2015, IBM produced a neural network chip, which it claimed had the computational capability of a mouse's brain. Human brains have nearly 1000 times as many neurons. So, if you ran 1000 of IBM's neural network chips in parallel you'd have the equivalent power of a human brain—and be able to point it at human problems. If you ramped up the power, you would eventually outperform even the best human beings.

IBM's Deep Blue did that against the world chess champion (Gary Kasparov) in 1996. IBM's Watson beat the best Jeopardy champions in 2011. Google's AlphaGo defeated Ke Jie the world's best Go player in 2017. And from about 2015 we witnessed the commercial dominance of AI, assisting corporations in the wake of the business craze for "Big Data" technology. You may believe it's the brilliance of the corporate executives that is forging the success of those companies, but it's not; it's the triumph of algorithms that can analyze vast mountains of data and use it to optimize business performance.

Intelligence can be defined as the ability to recognize real-world patterns and exploit those patterns through action. That computers can recognize faces and voices is not surprising. They can understand what you say and how to respond to your words as Amazon's Alexa does, for example. This powerful technology embedded in our digital devices may be useful. But the data it accumulates serves the giant corporations who provide it, just as the taxes imposed on the colonies served King George, rather than the interests of the colonists. The status quo will persist unless we change it.

The Distraction of the Cryptocurrency Mania

The technology that has the potential to upend the current status quo is the blockchain and we will use it to recapture control of our data. We will explain how and why this can happen in later chapters, but right now, if you discount the predictions of blockchain visionaries, it is not at all clear that any change in the status quo will occur. If you have followed the ups and downs of cryptocurrencies since they clawed their way into the business news, you might wonder "What exactly is revolutionary about cryptocurrency?"

Let's review the brief history of cryptocurrencies up to 2018 as a time line:

2009	Bitcoin created.
2011	Other cryptocurrencies, "alt coins" such as Litecoin, emerge.
2012	US crypto exchange, CoinBase, founded.
2014	The Mt Gox crypto exchange in Japan is hacked and suspends trading causing crisis of confidence in cryptocurrencies.
2015	Ethereum, the currently dominant crypto platform was launched.
2016	The DAO ICO raises over $34 million record amount—marking the start of the ICO craze which accelerates through 2017
2016	The DAO hack of Ethereum leads to Ethereum fork.
2017	US SEC intervenes in ICO market defining most cryptocurrencies as securities, stopping the US ICO market in its tracks.
2017	The Chicago Mercantile Exchange launches a futures contract in Bitcoin, signaling institutional acceptance of cryptocurrency investments.
2018	Massive rises in value of Bitcoin and other cryptocurrencies followed by a 80-90% collapse reveals Bitcoin to have been an investment bubble.

Table 2. The Bitcoin Time Line

The time line in *Table 2.* highlights a blockchain image problem. The public perception of blockchain technology follows the financial news: swings in Bitcoin value, large and successful Initial Coin Offerings (ICOs), fortunes made and lost, and so on. Few people take the trouble to understand blockchain technology and even if they did, it is not so easy to understand. The crowd of investors who have stepped into the cryptocurrency arena far outnumbers the aficionados who understand the underlying technology and its application. The financial bubble was a distraction.

Because blockchain technology can create digital currencies, the word on the street is that digital currencies may eventually threaten the banking sector. However, time has demonstrated that none of the current digital currencies could replace fiat currencies—they are too volatile, and none employed as day-to-day money. And beyond that there is little general appreciation of the other capabilities of the blockchain.

Do Currency Bubbles Happen?

Real currencies do not form speculative bubbles; commodities and securities do. The historical record is clear: all the famous historical bubbles—The Dutch Tulip mania (1637), the British South Sea Bubble (1720), the US stock-market bubble (1929), the US dot-com bubble (2000) and the US housing bubble (2007)—were bubbles in the value of commodities or securities, not of currencies.

Currency markets sometimes become volatile, but it is rare. When a currency becomes debased, from money printing (it's always money printing), the holders of that currency quickly dump it if they can. There are many historical examples of that phenomenon—the most recent ones being provided by Zimbabwe and Venezuela. But that's not a bubble situation where a dramatic increase in value is swiftly followed by a collapse. The holders of such currencies are citizens who dump the currency for one whose value is not volatile (often the US dollar).

Cryptocurrencies normally constrain the rate at which new "coins" can be created so they cannot be debased in the way that fiat currencies can. Nevertheless, they can suffer a collapse in value through speculative excess, and, in 2018, most of them did. As such they are commodities or securities.

A Consequence of the Lack of Digital Money

An important benefit of cryptocurrencies is that they enable micro-transactions. Currently, if you want to buy something on the Internet the cost of the transaction is somewhere in the region of 30 cents plus 2-3% of the amount of the payment, via a credit card or a payment service like PayPal. The cost was higher in the early days of the Internet and it is still prohibitively high in some situations.

Print Newspaper Advertising Revenue 1950 to 2011

Millions of 2011 Dollars

Source: Newspaper Association of America

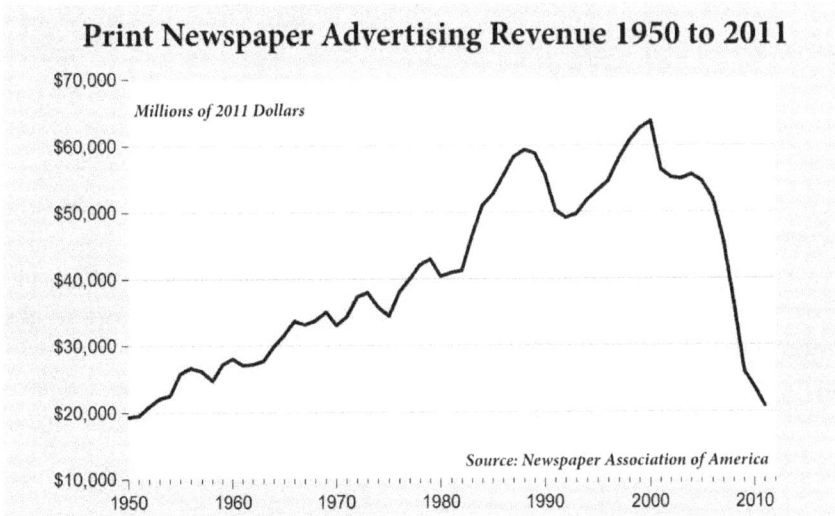

Figure 1. The Collapse in Newspaper Advertising Revenue

With such payment costs it is uneconomic to sell items costing less than a few dollars, except in multiples or as part of a large shopping basket. And that was fine for most eretail companies, but it didn't work at all for newspaper and magazine businesses. The Internet disrupted them, in many cases fatally, as can be gleaned from the above graph, which illustrates the catastrophic collapse of newspaper revenues.

Local newspapers lost the bulk of their classified ads, most of which emigrated to Craig's List. Websites were impossibly different—nothing like newspapers, with their distinct sections and stories (world news, national news, editorial, local news, entertainment, health, fashion, sports, business, comic strips, etc.). Web users never perused web sites as if they were newspapers. Aggregation sites like Yahoo News and Google News furnished links to all news stories everywhere; local, national or global. They invented a new and more efficient model of what a news service was.

The brand value of established newspapers and magazines was minimal on the web and subscription models worked for only a few. They had to depend on web advertising, so they made space for digital ads on their web pages and were paid cents by ad brokers when someone clicked on one. Payments were aggregated, with a single payment being made every week or month.

13

Digital advertising grew, with the introduction of middlemen, the ad-brokers, who took the lion's share of the ad revenue. The on-line ad market evolved into a technology contest between competing brokers who served up ads in real time. Newspapers and magazines lost control of their major revenue stream.

Every species of web publisher has to earn a living. A practical micropayments capability, would make a difference. Publishers and bloggers too, would be better able to directly charge for access to a single article or news story and the whole dynamic of web publishing might begin to shift. Web publishing could migrate to where a publisher can charge for the content and earn from advertising. Micropayments will have other consequences, which we will discuss later. However, the most dramatic, we believe, is the disruption of digital advertising.

Questions

You may have concluded from this first chapter, which touches lightly on what follows, that this book is woven from multiple strands. The initial question we tried to answer is this:

Does the blockchain herald a revolution, and if so, what will be the outcome?

When we examined the question, we concluded that it could indeed herald a revolution. So we sought parallels that would help us understand where the blockchain and its consequences might lead us all. Looking back in history, we found a technology and a revolution that came in its wake, that provided a parallel: the printing press and the American Revolution, which was promoted so effectively by Thomas Paine's Booklet, *Common Sense*.

In considering the outcome of the American Revolution, it became clear that it did not just install a new form of government, but stated and established what it saw as the "rights of man," which it enshrined in its constitution.

This led us to wonder whether this blockchain revolution will also establish new human rights. Specifically:

Will the blockchain revolution help us to establish our data rights?

In our efforts to answer these two questions we found it impossible to avoid investigating, analyzing and commenting on ten other questions, which we now list in the order in which we discuss them:

- In practical terms, exactly what is digital data?
- What should an individual's data rights be?
- What are the foundations of a currency?
- What are the characteristics of a viable currency?
- What are the characteristics of blockchain technology?
- What business impact is blockchain technology likely to have?
- What impact will cryptocurrency have on the advertising market?
- How is it possible to manage a cryptocurrency?
- How did the changes provoked by the invention of printing technology unfold?
- How might this new revolution unfold?

Immutable, Irresistible, Invincible

What we shall call blockchain revolution will not come to fruition because a few forward-looking thinkers decide it's a noble idea. It requires a change of attitude from millions, then hundreds of millions and then billions of people. They need to recognize that:

The ownership of personal data is a fundamental human right.

No debate is necessary; it's plain common sense. If people were asked to vote on this, who would vote against owning their data? It is your data and you have the right to control its usage.

Nevertheless, a change of attitude will have little impact unless the individual is able to control his or her own data and are also able to make productive use of it.

Here, the blockchain enters the equation. Blockchain technology enables the ownership and productive use of personal data. This will become clearer in later chapters. For the moment it is enough to appreciate that blockchain technology has already demonstrated that, without help from any bank or any government, it can establish a viable currency for peer-to-peer payments.

It can do so because *it can automate trust.* That's the superpower of this technology. It can be trusted with your personal data. And because of that it will change the world.

Tom Paine influenced the minds of two million colonists, helped to establish the rights of man and participated in the invention of a nation. His words were as consequential as Washington's leadership. He changed the world.

For the blockchain revolution, we need to influence the minds of two billion people and establish an individual's data rights. If we can do this, blockchain technology will change the world.

Chapter Summary

At the end of each chapter (except the last) we provide a summary of the main points or propositions contained in the chapter, as a reference for the reader. As this book spans multiple topics, these bullet-point summaries may help the reader keep hold of the different threads. For the final chapter, we provide a recapitulation of the primary points of the whole book.

The following summarize the main points made in this chapter:

- Recent news stories demonstrate that personal data is frequently violated, especially by social network sites.

- We have moved into a data economy, where the largest and most valuable companies are data miners, companies that have best managed to apply analytical techniques to very large accumulations of data—mainly personal data.

- None of the major digital currencies could replace a stable fiat currency, because their value is far too volatile. Historically there have never been any currency bubbles; there have been investment bubbles. The recent cryptocurrency bubble was just another investment bubble.

- The current digital advertising market grew partly because of the lack of digital money that was capable of making micropayments.

- One of the consequences was the damage done to the newspaper and magazine publishing market.

- Individuals have the right to the ownership and control of their personal data. It is a fundamental human right.

THE "COMMON SENSE" OF CRYPTOCURRENCY

Chapter 2

Differently Defined Data

"Such is the irresistible nature of truth that all it asks, and all it wants, is the liberty of appearing."

~ Tom Paine

———— ✸✸✸ ————

Before we broach the critical topic of data rights, we need to discuss what data is, and from there, determine what counts as personal data and what does not. Dictionary definitions of "data" are not particularly useful here. They make suggestions of the following kind:

- Facts and statistics collected together for reference or analysis
- Factual information used as a basis for reasoning
- Any representations to which meaning is or might be assigned
- The plural of datum

We can be much more precise, as we are focusing on digital data. If we cast our net as widely as possible, digital data includes any information that can be gathered or generated by any electronic device, including all computers, tablets and mobile phones. In short, all digitized data falls within our area of interest.

To ensure that data ownership is practical, we need to be able to store data securely, in a way that identifies who owns and who can have access to it. If you think data standards for this already exist, think again. They don't.

An Immutable Record

The early computer pioneers didn't do much thinking about data; they were motivated primarily by the applications (the software) that could manipulate data to create value. Looking in the rear-view mirror, with 20/20 hindsight, we might accuse those pioneers of a dereliction of duty, for this reason: They made an unfortunate and very consequential error when they introduced the idea of updating data.

Allow me to explain:

When you insert a new value, say an amount of $5.12, to replace an old value, say an amount of $12.10, you delete the old value you replaced. There is no longer any record of it.

In general, the problem is that an update destroys data, and consequently there is no longer any record of the previous value. In other words, there is no audit trail of what happened. The explanation is simple, but the implication is consequential. The absence of an audit trail means a fraudster can destroy or alter data, too, and leave no tracks.

To be fair, the people who made this error can be forgiven for doing so. Data storage used to be very expensive. Since 1960, it has fallen from the princely price of $10,000 per MB (million bytes) to the minuscule amount of $0.0001 per MB (million bytes). In the early days of computing, it wasn't affordable to keep a full audit trail of all data values. And as time passed, updating data became a programmer habit that persisted quietly by momentum.

This, by the way, is not a mistake that early bookkeepers made. On the contrary, the whole idea of bookkeeping was predicated on the principle that an audit trail was fundamental to the process. The postings to a ledger were immutable, just as the transactions posted to a blockchain are immutable. That is a critical aspect of blockchain technology. Aside from the other benefits it provides, it finally corrects an original sin of our digital ancestors which made computer fraud far easier to perpetrate.

Unselfconscious Data

Another unfortunate error was made when databases were invented. A database is a data store that is designed for sharing data between multiple programs, perhaps even hundreds of programs. Databases were designed for that purpose: to allow many different programs or users to share their data. Before databases, programs kept the definition of the data they were using to themselves and data was usually stored in files that contained no indication of what the data actually was. Only the program that used the files knew.

Database data is far easier for programmers or data users to understand, because databases provide a catalog of the data files they contain. The catalog defines what is in the files and the relationship between that data and other data files they contain. This development was a huge improvement on what went before, but sadly, a rigorous standard which designated how individual data records should self-describe and be identified was never created.

The consequence is this: Even data organized in databases does not know what it is or how it was created. It has no self-awareness.

The Self-aware Dollar

Rather than examining the technical intricacies and deficiencies of computer data, let's discuss a self-aware item of data that we are all familiar with: the one dollar note. Looking at the front face of the note, you are presented with a significant amount of information.

Figure 2. The Data on the Dollar

Think about it. The only information that you need about this note is what it is (i.e., US currency), the dollar amount (one), and whether it can be trusted. This is the same whether you are a US citizen, a foreigner, a business, a bank, or the Federal Reserve itself. And that information is right there, either printed on the note or enshrined in the design. It is a Federal Reserve note, as it announces at the top, and it says its value is one dollar.

It can be trusted, because a great deal of effort has gone into its design, making it almost impossible to forge. Aside from the watermarks, security thread, and the intricacy of the very high-resolution design—which, by the way, would require a very high-resolution printer to replicate—the dollar is printed using unusual ink which is difficult to acquire or imitate, and it is printed on special paper made from cotton and linen fibers, which is not available commercially.

In addition to that, the note bears a unique serial number, which gives the dollar bill a unique identity and includes the identity of the organization that created it, in the form of the Treasury seal and the signatures of Treasury officials.

To summarize, in addition to the data itself (the dollar amount), we have the following data:

- A unique identity (which can also be used to determine when the note was created)
- What it is (a single note of a currency of a nation)
- The creator (the US Treasury on behalf of a Federal Reserve bank).

You can look at a US driving license in a similar way. It can be trusted and is difficult to forge, given the mix of state seal, special inks, barcode, holographic stamps, etc. It has data (the details of the holder including a photograph), plus a unique identity, what it is (including expiration date and which types of vehicles it applies to), and the time of issue and creator (the DMV).

Other credential documents, such as birth certificates, wedding certificates, passports, green cards, etc., exhibit similar characteristics that, collectively, make those credentials self-aware.

From Self-aware Dollars to Self-aware Data

Digital data is not organized in the way a dollar is, but it could be and ultimately it will need to be. As soon as we raise the issue of data ownership, we are obliged to consider how every kind of data record can be stored and organized, so that it has a unique identifier, declares who created it, what it is, and who owns it.

You may be thinking that if we hold that kind of data for every little data record there is, it will take up a lot more data storage space. And it will, but data storage is cheap, and anyway there are reasons to believe that this could save data storage space.

The smallest amount of data, the atom of data if you like, is an event. Data is created when something occurs and the detail of what happened is recorded. Think of a simple example. Think of a sensor in an oil pipeline that does nothing but record a change in temperature. Maybe the temperature has to change by 1 degree C before it makes a report. That's about as small as a data record can be – an atom of data. There are many such small events. Someone clicks on a web link. Someone turns a light on. Someone starts their car.

We need to think in that way if we want to establish who owns specific items of data. Three things are required:

1. We need the data to declare what it is, when it was created, by which process and so on. Unless the data is specifically identified there is no possibility of determining who owns it.

2. We need the data to have ownership assigned to it when it was created, according to some sort of legal formulation. In the case of the oil pipeline sensor, the situation is simple: the owner of the pipeline owns the data. But in respect of your clicking on a web link, it may not be so simple. (We'll discuss this later.)

3. We need to have changes of data ownership recorded. This is common sense. One cannot sell or rent data one does not own.

If all digital data was organized in this way, we would find ourselves in a situation where individuals and organizations could trade their data.

Ten-tenths of the Law

Occasionally someone repeats the old saw that, "possession is nine-tenths of the Law." It is not true anymore. It was true in centuries past when a person in possession of an object had the strongest legal claim to it, whether that object was a mansion or a mandolin. And that was because there were few records to attest otherwise.

Consider this:

In medieval England, the ownership of land was conveyed in "fee simple," which, if you didn't know, means "permanent and absolute tenure," by what was called a "feoffment with livery of seisin." It worked like this: the seller and the buyer (also known as the feoffer and feoffee) met on the land that was being conveyed and performed a quaint little ceremony in front of witnesses. The seller would give the buyer a little symbol of the land itself, like a twig or a pebble or a handful of dirt, and then swore orally that he was passing the land to the feoffee. Once the feoffee moved onto the land, he possessed it.

This ceremony for transferring the ownership of land was eventually replaced by deeds — written legal instruments, which were signed, sealed, and delivered. In England, that never happened until the seventeenth century, which might seem rather late in the day. However, the truth was that for deeds, aside from paper and ink, you need people who can read and write.

It wasn't until the proliferation of print technology that literacy became an important skill. Printing enabled many things that were previously rare or unknown: bank notes, standard letters of credit, company stock, standard title deeds, and so on. Those financial instruments and deeds were just data on paper— officially standardized data objects.

You may not have considered this, but most of the valuable things you possess are provably yours only by virtue of such official data objects— data created by print technology which ruled the world before computers were invented and, in many ways, still rules it.

You may have thought that the 70 plus years of IT had completely usurped all of that Gutenberg technology. Nevertheless, we still have paper bank notes, paper letters of credit, paper stock and shares, and paper title deeds. These things will eventually be replaced by their digital equivalents, but it has not yet happened.

Data Protection and the OECD

Data rights have been a concern for the EU since 1980 when the Organization for Economic Co-operation and Development (OECD) produced recommendations about data protection. Headquartered in France, the OECD, had its genesis in the aftermath of WWII—it was formed to administer US and Canadian aid in the framework of the Marshall Plan.

It had considerable influence in respect of many European standardization initiatives. Encouraged by the OECD, the EU standardized its ideas on data protection. It began with legislation in France and the UK, and then, over a 30 year period, EU countries harmonized their data regulations. The culmination of this, the EU General Data Protection Regulation, known as the GDPR.

GDPR was implemented in May 2018, much to the surprise of most US businesses, who knew nothing about it. The surprise was twofold:

- GDPR legislation affected all businesses with even one EU citizen as a customer, irrespective of the location of the business.
- Violation of GDPR could lead to fines as large as 4% of annual revenues.

For advocates of personal data ownership, GDPR is a boon. There are now hundreds of millions of people who legally own their data. And it is likely that other democratic nations will soon adopt such rights. If a country asks its citizens to vote on whether they own their digital data, they will vote "yes." What started in the EU will gradually spread across the world.

The data rights that GDPR confers on EU citizens are:

1. **The right of consent.** No organization can store your data unless you freely give your informed and specific consent to every way that it uses the data, through deliberately opting in. Without that, no organization has the right to store your data, even if you willingly provided it.

2. **The right to be unpredictable.** The automated processing of personal data to analyze or predict an individual's behavior is curtailed. In particular, such prediction is forbidden if it will have a significant impact on the individual, such as in a hiring or credit decision.

3. **The right of access.** This stretches beyond having access to stored data. It includes: knowing what data is held; how it was acquired; how long it will be kept; how it has been processed and why; to whom the data has been disclosed; and if it has, how it has been protected.

4. **The right to change.** You have the right to change (i.e., correct) any of the data held about you.

5. **The right to erasure.** This is also called the "right to be forgotten." No organization can hold your data if you do not want them to. They must delete it.

6. **The right to portability.** You can request all your data from an organization, and it must be transferred in a "machine-readable" format—such as a CSV file.

7. **The right to know.** You have the right to know whether an

organization (any organization) is holding your data. It must tell you.

8. **The right to complain.** And finally, the "don't mess with my data or I'll report you" clause. There is a supervisory authority (the EU data police) to whom you can report data misdemeanors and felonies.

The Universe of Personal Data

GDPR defines personal data in the following way:

Personal data means any information relating to an identified or identifiable natural person (a data subject). An identifiable natural person is one who can be identified, directly or indirectly, in particular by reference to an identifier such as a name, an identification number, location data, an online identifier or to one or more factors specific to the physical, physiological, genetic, mental, economic, cultural or social identity of that natural person.

That's a reasonable leaping off point, but we can assemble a more comprehensive picture by assigning specific categories to personal data, as follows:

Basic personal information

This is the kind of data that you might be asked for when you sign up on a website. It comprises basic identity data, such as name, gender, date of birth, and general location data such as zip code, city, and state (or the equivalent according to country).

Personal contact data

This is data that enables people or businesses to contact you. Thus it includes your full address (or addresses), telephone number(s), email address(es) and any other IDs for messaging applications. It may also include social media IDs, particularly your LinkedIn or Facebook ID, and maybe even profile photographs, since when coupled with your name they can be used to try to find you on the Internet.

Personal documents and credentials

These are formal identification items that have become credentials that help you prove who you are. They vary from country to country and include such things as driver's license, work visa, passport, birth certificate, marriage license, social security number, and so on. These are credentials from some issuing authority (usually some department of government) that verify a person's identity.

Personal digital credentials

This category includes all digital keys that provide you with access rights, including access to websites, software applications, computers, or wallets for digital currencies. Conceptually, these could be added to the category above—they just happen to be digital.

Personal history

This is your personal history. It includes previous addresses and contact records, your education record and associated documentation, your employment record, and all travel. It also includes a record of skills, courses completed, professional qualifications, languages spoken, and so on. It includes your criminal record if you have one, and any interaction you have had with the law, whether criminal or civil.

It would naturally include all your family information: parents, siblings, children, marriages, divorces. It includes your family tree as far back as it can be verified.

Health information.

Data such as vaccine history, medical history, doctors' reports, blood type, lab results, scans, dental records, medical insurance, genetic data, etc., are in this category. It includes your health data from all your health care providers. We separate this from personal history for two reasons:

1. It includes information that might be required in an emergency.

2. It is data that is regulated in many countries.

Financial information.

Like health information, personal financial information is already subject to regulations which can vary from country to country. It is the data relating to bank accounts, debit cards, credit cards, crypto wallets, stocks and other investments, and other similar financial assets and liabilities, such as mortgages and loans.

It also includes a full record of all tax information and tax payments and the full record of all purchases and sales of any kind that an individual has made, as well as details of all business arrangements (accountant, brokers, lawyers, etc.).

Title

This category focuses on possessions that have value and the proof of ownership of such possessions. The data here includes deeds, titles, provenance, appraisals and other documents that relate to or prove an individual's ownership of possessions such as a house, car, antiques, etc. It also includes any items that you have created or manufactured, machines, works of art, books written, music, videos, and any patents.

Memberships, Interests, and Preferences

This category includes data or documents related to membership in associations, societies, and groups of any kind, such as sports clubs, retail warehouses and so on. Membership may imply specific interests and activities, but people may have many such interests for which they have never joined any society. People have political and religious interests that may not be recorded or documented in any way and which may change with time. All such interests and preferences fall into this category.

Digital possessions

This category is for all an individual's digital possessions, including all photographs, videos, music, sound recordings, the software they have bought, data files, and so on. It also includes any data created by you and digital devices you own and use.

The Personal Audit Trail

This is a person's digital history. It comprises an audit trail of an individual's digital activities and interactions, in effect their Internet history plus the history of any activities carried out on a computer or a mobile device. Given that, for example, a mobile phone can track your location, this category could also include a complete timeline showing your personal location.

Deduced data

As you can see, if you take everything into account, personal data in all its glory is incredibly diverse. One of our motivations for analyzing personal data in this way was to see what issues surfaced in the course of listing them. An important issue it surfaced was the problem of *deduced data*.

This category of data is created as a result of processing other personal data. For example, you might want to use some of your personal data to calculate how much you spent on average every week on foods of various kinds. There are probably many ways you could use personal data to discover useful information, and of course the big data businesses have been doing that with your data for years.

The Deduced Data Conundrum

Personal data exploitation is not the result of a massive conspiracy. It wasn't as though a group of super villains got together and plotted to steal people's data from under their noses and squeeze a fortune out of it. Companies like Google, Facebook, LinkedIn, Equifax, and so on, all had business ideas (ad-based search, ad-based social network, business network, credit scoring) and they gathered the data they needed to pursue their businesses.

They hired smart guys, and the smart guys found ways to improve the profit that these businesses could make. And if any of the improvements involved gathering more personal data, then naturally — as there was no law against it — they did exactly that. They never assembled anything like the full inventory of personal data that we described. They just grabbed what their business could use — and, of course, some of them sold your data to others.

It was your data, supplemented at times with data they deduced about you, that you probably never knew existed. When a company is in the business of harvesting the personal data of millions of people, it can throw gangs of algorithms at it and discover exploitable needles hidden within the giant haystack.

Perhaps the greatest area of concern involves the application of psychometric algorithms that can measure and predict people's behavior. One particular example of the application of psychometrics illustrates this concern. An expert in the field, Michal Kosinski, was behind the psychometric analytics deployed by the Trump campaign, courtesy of Cambridge Analytica, to influence the 2016 election—using your data from Facebook.

The Facebook Psychographic Data Story

In 2012 Kosinski demonstrated something surprising. By analyzing an average of just 68 Facebook "likes" of one individual, it was possible to predict their skin color (95% accurate), their sexual orientation (88% accurate), and their affiliation to US political parties (85% accurate). It was also possible to predict intelligence, religious affiliation, and alcohol, cigarette, and drug use, with high levels of accuracy. It was even possible to deduce whether someone's parents were divorced.

Here is how this works. The data analyst builds psychographic profiles based on your preferences, and these profiles are used to predict inclinations, opinions and facts about the individual. Psychographic modeling data is data that most people do not know exists. Few people even know that this field of activity exists. It is predictive data, and its predictions are disturbingly accurate. And if such data can be useful to the big data brokers, it is quite likely that it can be useful to you.

Can We Be Anonymous?

Many of the services you may someday want to pay for require that you share your data with the business providing the service. If someone is going to make you a suit, they will need to take measurements. If you want to buy life insurance, the insurance company needs some basic details and maybe also some of your health data.

An interesting question is how often does any such business need to know who you are? Most businesses do not need to know. Your tailor doesn't need to know, neither does your grocer, or the hotel you stay at, or even the airline you fly with. They may gather your personal data, but they don't necessarily need it.

Consider the act of paying for something. The only thing the business you are paying needs is the correct amount of money to be deposited in their bank account. They may need a unique ID reference to store with the payment, but it doesn't need to be your name. It could simply be a unique anonymous reference.

In many situations, it is possible to separate your personal identification data from the data required for the business interaction. Consider a very personal scenario—a medical emergency. Both the doctors and the patient will want all the needed personal medical and insurance data to be available, but they still do not need your identification data.

If you are wondering why this matters it is because data is easy to copy. It is so easy to copy that it is difficult to imagine any practical way of preventing data from being copied. An organization may legitimately need to view a small portion of your data. But if it never gets access to your personal identification data, it cannot easily build a collection of your data, or pass your data to others to build up such a collection.

This suggests a best practice for personal data:

Unless it is impractical, only provide access to anonymized data.

Chapter Summary

- Currently, computer applications do not include audit trails of their transactions. This is partly because of an error made in the early years of computing, allowing data to be updated. (Updating data destroys the previous value.)

- EU General Data Protection Regulation (GDPR) confers data rights on European citizens. They are: The right of consent. The right to be unpredictable. The right of access. The right to change. The right to erasure. The right to portability. The right to know.

- Because of the way GDPR has been implemented, it applies to companies world-wide who have even just one EU citizen as a customer.

- We can classify personal data into twelve categories, according to the following scheme:
  > Basic personal information
  > Personal contact data
  > Personal documents and credentials
  > Personal digital credentials
  > Personal history
  > Health information
  > Financial information
  > Title
  > Memberships, Interests, and Preferences
  > Digital possessions
  > The Personal Audit Trail
  > Deduced data

- Deduced data presents a specific difficulty because, even though it is personal data, it may be created mathematically through aggregation and data analysis. The power of psychometric analysis ought to be particularly worrying.

- Technically, it is possible, in most circumstances, to build applications that allow data to be used anonymously. In many situations businesses do not need to know your personal details, they only need to know how to contact you.

Chapter 3

The Data Rights of Man

"No price is too high to pay for the privilege of
owning yourself."

~ *Friedrich Nietzsche*

———&—

There is no easy way to place a value on personal data aside from digital objects (music, videos) and the title to physical goods. Nevertheless, there are four areas where we can estimate data value:

– digital advertising

– stolen data

– a credit score

– an individual's data tracks

We will discuss these different views of data value one by one.

Your data value as an advertising target

Let's consider the price advertisers pay to try to catch your eye. Facebook, Google, and other digital ad brokers use your data for targeting. They capture your behavior and your consumer profile (your preferences, lifestyle, stage of life and various other attributes).

Facebook prospers from the fact that it knows a great deal about its users—its average US user spends about 40 minutes a day on the site, enhancing that knowledge. Facebook regularly runs batteries of statistical algorithms to match users with the products advertisers wish to promote. In 2017, the average cost per click for an online Facebook ad was $1.72—a premium price that stems from the mountainous variety of data it can dissect and evaluate.

Google, the other giant of the digital domain, cannot match Facebook in this area, but more than compensates for it with Google Adwords, which, with an 80% share of the US market, dominates search advertising. According to Wordstream, in 2017, the average cost per Google AdWords click was $2.32. Naturally, some clicks cost more than that. The price is fixed by auction, so it varies with demand. The price for AdWords for legal services, for example, can rise above $50 per click. Nice work if you can get it.

If we take the total US revenues from digital advertising in 2017, about $83 billion, and divide it by the population of US Internet users (roughly 287 million), you get the digital ad revenue per average Joe or Jane. It works out to be $289.19 per annum. It is an average, so if you do many product searches, or frequently click on website ads, your total will be higher—especially if you often seek legal services.

Your data value to hackers

Another way to look at the value of personal data is from the thief's perspective. Data thieves usually steal personal data so they can sell it on the Dark Web to other thieves. If it wasn't worth much they wouldn't bother. The bare details of a credit card (name, card number, expiry date) have little value. But if you add in the owner's address and email, then it's worth somewhere between $20-$25. It has a similar market value to a driver's license. So one debit card, two credit cards, and a driver's license, plus your email and physical address commands a price of $100.

Passwords can be valuable. Your Netflix password is worth about $3.00. Your Spotify password comes in at about $2.80. A password that walks you into a bank account with a balance in the region of $2000 commands a price of $100. For a balance of $15,000 or more, think in terms of $1000. A complete medical record can fetch the same $1000 price, although, like the bank account, the value depends on what it contains. Such details can be sold to insurance companies, or even used for blackmail, but the less it includes, the less value it will command.

Estimating an average value for passwords and personal credentials is difficult. But if you include a collection of passwords for a bank account, a savings account, add in a few credit or debit cards, a driving license, and a passport and assume just an average medical record, at current prices, it's in the region of $300.

Your value as a credit score

In 2016, Equifax made a healthy gross profit on revenues of just over $3.1 billion—the year before it managed to compromise the personal financial data of 147 million Americans. Credit scoring is a profitable business, as both Experian (annual revenues of $4.55 Bn) and TransUnion (annual revenues of $1.7 Bn) can attest.

Roughly a quarter of those global revenues flow from the tens of thousands of companies that are interested in the creditworthiness of about 235 million Americans. The data that Equifax, Experian, and TransUnion present to those companies is your personal financial data, gathered, aggregated and analyzed without so much as a "by you leave." Do the math and you'll discover that they make about $10 per annum from the average Joe or Jane.

The value of your data tracks

Last, but by no means least, are your personal digital tracks—the full history of your digital activities.

To get some idea of the value of your digital tracks, consider an experiment conducted by Federico Zannier, an alumnus of New York University and an experienced IT consultant.

Zannier decided to sell his personal digital tracks for $2 per day over one month, using Kickstarter. The data he included were: the text of every web page he visited, regular screenshots of his PC activity with timestamps, a folder of webcam photos taken every 30 seconds, a log of all PC application activity (open and close times), browser activity including searches, personal geolocation, and PC mouse movements.

Federico guessed he'd earn about $500 from his one-month data sale, but exceeded that target more than fivefold. He raked in $2733!!

As with all other categories of data we have already discussed, different people would command different prices for their digital tracks. The digital tracks of an A-list celebrity would surely command a higher rate than Federico's; those of the average Joe or Jane, far less. Nevertheless, they are probably worth about $1000 per annum.

The Inability To Monetize

To determine an accurate average value for US personal data would require much more research than we have done here. Nevertheless, we can hazard a guess. Given the data we have discussed and the extensive nature of an individual's data resource, we tentatively suggest that, on average, the personal data of a US resident is worth somewhere in the region of

$2000—$3000 per year.

However, this money cannot be earned; it is merely potential revenue that could be realized if we lived in a true data economy. We can only speculate on the possibilities for data trading when a true data economy does emerge, but we suspect it would double the figure we gave above.

In the Direction of Data Rights

Let's now examine what the data rights of man ought to be. We will begin with two easily understood propositions:

1. Data is property. Your personal data is your property.

2. Your personal data is subject to your copyright.

The advantage of these propositions is that centuries of established law and precedent cover property and copyright, so we do not need to discuss what these propositions mean in most circumstances.

Let's imagine a world where all data is formally owned and securely stored. Let us assume it has happened; all data objects know what they are and know who or what owns them. Their ownership is determined by who holds the private key that provides access and control of the data.

So what happens when a new person is born?

The Birth of a Data Owner

Your data life begins at conception, and inevitably includes family data (about parents, grandparents, etc.). Concerning your health, your medical records are intermingled with your mother's medical records, and she may share some or all of that data with you, but (unless some court makes a ruling on this) she has no legal obligation to do so. Once you are out of the womb, all your medical data is yours, and other data will gradually accumulate. It is yours only in the sense that it will eventually become yours. For many years (exactly how many will probably vary from country to country) your parents (or in the absence of parents, your appointed guardian) will have control of it.

There are issues here that will need addressing. Consider, for example, if your father is not your biological father, do you have a right to know? This is not a simple issue. If you have the DNA record of your father and mother, you can deduce such information that either or both would have preferred to remain hidden—but do you have a right to access the data?

Your parents are custodians of your data until you reach the age of "data majority" after which it's your data, and they no longer have access unless you provide them with it. However, life is not simple. As an adolescent, in respect to some of your data, you will probably want sole control of access. As a child, you may receive gifts or earn money or receive an allowance, and you may acquire digital possessions, all of which you should have the right to control. Rules will be required. We can suggest some:

- Personal data must never be destroyed.
- Personal data may have a custodian who controls its use on behalf of another (e. g. parent for child) for a defined period of time.
- At some legally defined age, an individual is introduced to the digital data world. From that time on they have control of some of their data.
- At a later legally defined age, an individual takes full control of their data from its custodian(s).

The need for a data custodian may also occur in respect of someone who is severely disabled or incompetent to manage their own affairs.

Fundamental (National) Credentials

The fundamental data credential is the birth certificate. If you don't have one, life in most countries becomes difficult if not impossible. In the US, the birth certificate (from any country) is required to get a Social Security Number and a Driving License, which are the practical credentials (together or separately) for most situations anyone is likely to encounter. In most countries there is a central register of new births and the registering organization holds the master copy of the Birth Certificate.

The situation in the US will seem strange if you are unfamiliar with it. All birth certificate copies include a serial issued on special Bank Bond paper and authorized by "The American Bank Note Company." This dates back to 1913 when the US introduced the Federal Reserve Act. This established The Federal Reserve Bank as a private central bank whose role was to regulate the amount of money the US government was allowed to borrow and put in circulation.

This central bank arrangement worked fine until the Great Depression. In 1933 the US was unable to pay its debt and was effectively bankrupt. The private banks that made up the Federal Reserve demanded their money and the government of FDR, unable to pay, assigned all US citizens to be collateral for the debt. By that mechanism, the property and assets of every living U.S. Citizen were, and still are, pledged as collateral for the US National Debt. For US citizens, technically, the validity of your identity is predicated on you accepting your part of the national debt.

In every country, the practical reality is that some government agency registers your identity and vouches for it. Currently, your identity (name, date of birth, parentage, location at birth) are clearly yours, but they may not be wholly owned by you either. For all practical purposes that ownership is shared with the government, because it (and currently only it) can validate that information. Because of that, in the US, the government has assumed the ability to use your worldly goods as collateral for its debts—in other countries the situation may be similar.

Derivative Credentials

All other credentials, educational (qualifications), professional (qualifications), financial (credit scores, etc.) are to some extent jointly owned. The word "credential" derives from the Latin verb "credere" meaning "to believe," and the other party to the credential vouches for that belief. This is a valuable service without which an economy could not function.

We have become so accustomed to it that we rarely notice it. We assume that our doctors, dentists, accountants, bankers, lawyers and so on are qualified, because credentials exist to prove it and we presume they are kosher. We assume the airline pilot has had the requisite training. We presume the Uber driver who picks us up is legit. We expect the authors of our school textbooks are well chosen. All of these situations involve regulatory processes, credentials and shared personal data. And the associated data is data we wish to be accessible rather than hidden.

The Death of a Data Owner

Once someone achieves "the age of data majority," and their data comes under their control, in practice it comes under the control of a private key. We will explain the technology later, but in simple, practical terms the private key is a single access password which is unique and irrecoverable.

Someone could die, and their will could assign the ownership of all their personal data to, say, their favorite niece. However, the beloved niece will not be able to get her hands on that data unless she gets her hands on the private key. If, for whatever reason, the key cannot be found, the data is irretrievably lost. For most personal data it will just have to be accepted, but for the title data that proves ownership of physical things (land, houses, furniture, etc.) its irretrievable loss needs to be prevented. A second copy of such data needs to be held under the control of a custodian who will relinquish its control only under the directions of an appropriate authority.

Most likely the data will not be lost. In which case, some of the data will need to be tagged to indicate that the owner is dead. Credentials like a driving license, or some professional qualification, may have historical interest, but they have ceased to confer the right to drive a car, or to apply for a particular job. Local tax law will determine the tax on the digital estate.

Unreasonable Search

All governments, no matter where, levy taxes and treat tax evasion as a crime. No doubt they will want to have access to a citizen's personal financial data. But they can only have such access if the data owner provides it. The question then is what right does an individual have to deny access to their data to agents of government. The fourth amendment to the US constitution reads as follows:

> *The right of the people to be secure in their persons, houses, papers, and effects, against unreasonable searches and seizures, shall not be violated, and no warrants shall issue, but upon probable cause, supported by oath or affirmation, and particularly describing the place to be searched, and the persons or things to be seized.*

We can apply the fourth amendment literally to personal data, and we should. The government has no right of access to personal data without probable cause. It would be "unreasonable search." And even where there is probable cause, if your personal data is secured by a private key, it won't matter if they turn up with a SWAT team, the only way they will get to search the data is if you give them access.

The law is the law, so if you refuse to provide access in the face of a valid warrant, you will suffer the consequences. But there's a twist. The powers that be will not be able to get a warrant to search a data store unless they can prove it exists. There are ways to create data stores that cannot be traced, as we shall explain when we discuss zero-knowledge proofs.

Some Data is Born Shared

In most respects, the laws regarding the shared ownership of anything were created a long time ago, and there is no reason to expect them to be much different if what is being shared is data. However, there are some specific situations that are worthy of discussion.

Some data is born shared. For example, you send someone an email. It's your personal data (you created it) and by the act of sending it, you agreed to share the data with someone else. Communication is, after all, the act of sharing data, so all communication data is of this ilk. Once it becomes a conversation, then it is logically jointly owned between the participants, unless some specific agreement about data ownership has been made.

Under most circumstances there is no implicit privacy contract involved with such data and, unless you attach privacy rules, the other party may have the right to share it with anyone. Nevertheless, the ownership of such data is truly shared, so whatever regulations apply to changes to its usage it would require agreement between the owners, just as if it were a physical object.

Social media postings are shared data of a different kind—they are personal data that an individual has deliberately chosen to share. There is an explicit agreement that others can view the data you put on those sites, which you may be able to control. In most cases, this is data that you want to share. In some cases, it is data that you are happy to share with everyone in the world. There is a difference between providing access (you can view my data) and allowing data to be copied (you can take my data and process it in any way you please). If you allow data to be copied then a set of usage rules ought to be set.

Data Copyright

It is difficult to argue against the proposition that your personal data should be covered by a data copyright—and that anyone who uses the data without your permission violates the copyright. However for that to be the case the current definition of copyright needs to be extended. Copyright protects the intentional creation of specific items: works of authorship and artistic works. But it does not protect facts and it will need to, including accidentally created personal data. It doesn't protect simple things like recipes, unless accompanied by substantial literary expression in the form of an explanation or directions.

There is the possibility of data duplication which needs to be considered. For example, I create a recipe for scrambled eggs, simply noting down how I do it one morning, and it turns out to be just the same as someone else's recipe. Neither of us copied from the other. Who owns the data?

It doesn't matter if both of us keep that data private. Neither does it matter if we both publish it. It only matters if one of us finds a way to monetize that data—at that point digital rights enter the picture. Digital rights have been a thorny issue since the birth of the Internet. People steal data, whether it's ebooks or music or video. It possible, so it happens.

To create order in a world where people own their personal data, all data theft has to be stopped in its tracks. It doesn't matter that for decades people have been stealing pop music. In the new world it has to be prevented. We have to insist on principle that people have digital rights over their data and that copying their data without permission violates those rights. We will discuss how to achieve this in later chapters.

Data Aggregation

We drew attention to the issue of *deduced data* in the last chapter. Businesses that gather personal data can create such data, and its data that you do not necessarily know about. There are two issues here.

1. Is it right for an organization to aggregate data and create personal data about you without sharing that data with you? If you gave them permission, of course they have the right to do such analyses. But if you knew what they might discover and what value they may derive from it, then maybe you would think twice before allowing it.

2. Perhaps data owners themselves ought to collaborate in aggregating and analyzing data for their collective and individual benefit.

Three Types of Data

In the previous chapter we divided data into different categories (twelve in total). However, from a practical control and management perspective, we can divide it into just three categories, as follows:

1. **Credential Data.** These are data objects such as birth certificates, driving licenses, credit cards and so on. There are two kinds of credential data: primary credential data and secondary credential

data. The primary credential data is the birth certificate vouched for by a recognized validator such as a government. Secondary credential data can only be awarded by a validator who first validates the primary credential data.

2. **History Data and State Data.** These are data objects or data collections which record the activities and states of the data owner. It could be medical data, educational data, financial activity, even website activity or your location at a specific time. This is data about you.

3. **Title Data and Data Possessions.** Title data is data that proves ownership of some physical thing. A data possession is something that exists entirely as data (like a crypto wallet or a digital photograph). These are possessions, and thus their ownership can be transferred.

Armed with these three categories, we can propose a "Declaration of Data Rights." So let's do that.

A Declaration of Data Rights

1. **The foundation of an individual's personal data is their identity, which is fundamental credential data. It is their personal property vouched for by a recognized validator—in most instances conferred at birth by a national government. These credentials cannot be revoked.**

Any additions to fundamental credential data are subject to verification by the original validator or its delegates.

In vouching for the individual, the validator may also confer other data on the individual such as citizenship and other rights which that confers. The validator holds its own copy of this validation data to use for validation purposes. This data can never be revoked.

2. **Fundamental credential data provides the foundation for all personal data, as all other personal data links to it. The validator guarantees to provide validation in perpetuity.**

This is to cover any fundamental credential change as could happen with a change of validator (a change of citizenship is an example of this), or change of gender or change of name. Such changes are recorded as additions to the data. The original data is never changed.

3. **Secondary credential data may be provided by any validator who validates the individual's identity in the process of providing the credentials. Such data may be subject to contract.**

This covers any credentials of any kind, from important credentials, such as a driving license, to less important ones such as a club membership. Such credentials, like a driving license, can be subject to a formal contract and may expire.

4. **The management of personal data by those incapable of managing it themselves is through a custodian or joint custodians. Incapacity can be due to age (as in the case of children) and admits three possibilities: fully incapable, capable of managing data privacy and communications only and fully data capable. Jointly owned data that is not subject to a formal contract is equally shared among its owners.**

This is to cater for the custodial management of personal data. It differs from other property only in respect of the right to privacy for personal communications. Custodial management would typically be limited by time, so that, at a particular age, a minor would become fully capable of managing their own data.

5. **Explicit permission is required for the recording and retention of personal history data and state data.**

It will be necessary for systems reasons to be able to track a person through your web site. However, the data can only be retained if permission is granted.

6. **Personal history data and state data is not transferable. It can only be rented.**

This is designed specifically to prevent anyone from acquiring ownership of the personal data of others. No doubt, if it were allowed, someone would try to gain control of that data.

7. **Title data in all its forms is fully transferable from one owner to another.**

Note that cryptocurrency itself is simply a digital possession, as much as a movie or a pop song.

8. **Personal data is property, subject to both property and copyright law.**

Data is unusual in that it can be stolen by copying as well as by direct theft. The violation of copyright by copying personal data without permission is thus theft. This is the case even when personal data, for whatever reason, is in an unprotected copyable state.

9. **Personal data has the right to be secure against unreasonable searches and seizures and it shall not be violated. No warrants shall be issued for access to it except upon probable cause, supported by oath or affirmation, and particularly describing the data to be accessed and searched, and the data to be seized.**

This is the fourth amendment to the US constitution, repurposed to apply to data.

10. An individual's personal data cannot be used in evidence against him or her without the individual's permission.

This is the fifth amendment to the US constitution, repurposed to apply to personal data.

Chapter Summary

- The value of your data can be estimated in different ways. In this chapter we examined four:

  > Its value to a digital advertiser for ad-targeting purposes
  > Its value to a data thief
  > Its value when used to calculate a credit score
  > The value of your data tracks when sold.

- We estimate the personal data of a US resident to be worth, on average, somewhere in the region of $2000—$3000 per year.

- In respect of data ownership, there is an issue as to how to handle the custodianship of data of someone who is too young to manage their own data.

- There is similarly an issue in disposing of data when someone dies.

- There is an issue in terms of the handling of data that is jointly owned.

- Personal data can and should be covered by copyright.

- Individuals could, if they collaborated, take advantage of data mining algorithms that are currently used against them.

- Distinct from the 12 categories of data that we described in the previous chapter, in this chapter we suggest three distinct types of data, based on how the data can be used. They are: Credential Data, History Data and State Data, and Title Data and Data Possessions. Credential data is data that needs to be used to validate your identity or entitlements, History and State Data is simply data about you that you own and of which you should never be able to lose control, Title Data and Data Possessions are tradable data whose ownership can be transferred.

- The chapter concludes with a proposed declaration of data rights.

Chapter 4

Cigarette Money and Other Currencies

What we obtain too cheap, we esteem too lightly; it is
dearness only that gives everything its value.

~ Tom Paine

———

Let's discuss money. It is difficult to imagine an economy, even a small one, where there is no money, and all trade is carried out through some form of barter. There are a few documented examples. The Yanomami and also the Awa tribes of the Amazon rainforest are nomadic hunter-gatherers who have survived for as long as anyone knows without any form of cash. Neither is there any kind of bartering; they just live co-operatively. In slightly more complex environments, for instance, among the Nyimang people, wealth and status are determined by how much cattle you own. They trade their cows with other tribes occasionally for seeds. You could claim it is barter, but realistically, the livestock are just money.

That's the general tendency among all societies. Whatever is the interchangeable or if you prefer, fungible, becomes money. At various times food (grains, rice, tea, salt, peppercorns, fish, meat) has served as money. So have ornamental things: trinkets, beads and cowrie shells, and useful things, like blankets. A means of exchanging items is necessary for even relatively simple societies, so money of some kind emerges.

Post-WWII Germany provides an interesting case study of an unusual currency that worked effectively in a large economy. The second world war completely wrecked the German economy, and the country was split into four zones under the military rule of the Americans, British, French, and Russians. Hitler's Reichsmark was still in circulation but had been rendered almost worthless by massive inflation in the latter stages of the war. And because Reichsmark notes bore the swastika, the Western administration withdrew them from circulation, replacing them with Allied Occupation Marks, printed in the US.

Getting this temporary new currency to circulate proved difficult—so difficult in fact that a different kind of money emerged instead and became

47

dominant: cigarettes. This unusual development happened almost naturally. In Germany smokers' ration cards had been issued from 1940 onwards, limiting smokers to a mere 40 cigarettes per month. Since most smokers have a more substantial appetite than that, a black market for cigarettes soon flourished. When the war ended, cigarettes were the most effective currency because they were already circulating, the supply was limited, they had value and they were regularly burned—damping down inflation.

In June 1948, the Deutsche Mark was launched, with an "airdrop" of 40 DM to every resident of West Germany. At last there was a widely distributed and reliable currency in circulation which quickly displaced cigarette money and forced the black market into retreat. While cigarettes were obviously not a sustainable currency, they proved to be effective for a few years among a population of over 30 million.

From the mists of time

By definition a currency is a medium of exchange—it must be exchangeable for goods. Until the dawn of the digital world, this severely limited what could serve as a currency. It couldn't be something that decayed easily, or could be forged easily, or was too large or too heavy and, above and beyond those practical details, people who used it needed to have faith in it.

These conditions considerably diminish the possibilities of what can serve as a currency. Historically, until the invention of printing, there are few examples of money that was not coins of one kind or another, made from iron, bronze, silver or gold. Thousands of years of unrecorded history obscure the origin of the metal-based money. It may date as far back to the earliest evidence of metal working around 8,700 BCE. Money, in the form of precious metals, existed in ancient Babylon and Egypt, where standard shapes and weights were money. Greek historian, Herodotus, claimed that coins were invented in Lydia (now modern Turkey) in about 700 BCE and the archaeological evidence agrees.

The invention of "the account" is similarly lost in history. "Account money" mandates a trusted organization that will hold and manage the safety of people's money and keep an account of who is owed what. Such money existed in ancient Egypt, Babylon, India, and China, where temples and palaces often included commodity warehouses that issued "certificates of deposit" as a claim on the valuable goods stored there.

However, the storing of valuable possessions (account money of a kind) might go back earlier than that, depending on what you think "tally bones" were used for. The "Wolf bone," discovered in 1937 in Czechoslovakia during excavations at Vestonice, Moravia, has been dated to about 30,000 years ago. The bone bears 55 marks, which are believed to be tally marks—although it is impossible to know what they tally. Another "tally bone," the Ishago bone, discovered in 1960 in the Belgian Congo, has been dated to 18,000 to 20,000 BCE. It has a series of tally marks carved in three columns.

Bearer money and Account money

The distinction between **bearer money** (coinage and notes which belong to whoever holds them) and **account money** (a managed store of value) is a fundamental one. These two types of money are complementary and have existed for millennia. If we consider a currency to be something that exists in both these forms, then in order to be effective it will need to have most or all of the following characteristics:

- It can be exchanged for goods or services, as the medium of exchange.
- It is a metric of value in the sense that people used it to measure the value of goods and services. (Consequently, its value must be stable.)
- It is a store of value, and consequently can be used as a currency of account (where an appropriate accounting mechanism exists).
- It is easily divisible.
- It is inexpensive to manufacture.
- It cannot easily be forged.
- It is highly portable.
- It is widely accepted (or is formally designated as legal tender).

Declaring a currency to be "legal tender" can be a government tactic to enforce the use of a currency. It forces all merchants to accept the currency as money, whether they wish to or not. Naturally, such regulations only apply within a given jurisdiction. However, if the users of a currency have confidence in its value, such a regulation is unnecessary.

We will have more to say about these saintly currency characteristics later. But first, we'll review the history of money. We can begin by discussing coins—the once dominant *bearer money*.

The Archimedes Displacement

When King Hiero of Syracuse became suspicious as to whether a golden votive crown he had commissioned from a local goldsmith was solid gold or not, he asked Archimedes to find out and to do so without damaging the crown. Archimedes did not know how to do that but invented the solution when stepping into a bath. His subsequent naked dash through Syracuse shouting "Eureka!" is legendary. That happened around 250 BCE and it marked a significant change in the technology of money. There had been no known way to prevent the debasing of precious metals, and now there was.

It only partly solved the problem. You could use it for standardized ingots and for collections of coins, but it was not accurate enough for single coins. They were too small to test effectively.

Coins may appear to be an effective medium of exchange, but in practice a coinage economy is fraught with problems. The typical approach to minting a coin is to create one that includes a defined proportion of a precious metal (gold or silver) and some other metals (copper, tin, nickel etc.). The value of the coin is pegged by the value of its precious metal content, so the coin is worth slightly more than the cost of mining the metal. The government controls the mint and thus makes a profit from every coin that it makes.

So what are the problems?

The Challenges of Coinage

Consider the Roman Empire which lasted a thousand years. The Romans had a multitude of different coins of different relative values. In the time of Augustus (around 27 BCE) there were nine, the largest being the Aureus—worth 1600 Quadrans, with the Quadran being the smallest coin. They all had precious metal content - mainly silver.

The Italian peninsula didn't have any large reliable mines and thus the mint had to get its precious metals from elsewhere. It could acquire some by conquest (the spoils of war), and it could also demand tribute money and taxes from newly-conquered lands. Conquest was the economic engine of Rome for centuries. When Rome ceased to expand, as it eventually did, it leaned more heavily on the mines of Greece and Spain for its coinage. The Romans also suffered a loss of coinage through trade along the Silk Road, particularly with India—nowadays we would call it a balance of payments deficit.

The Romans' solution to his problem was to melt down older coins to produce newer coins containing less precious metals. Just as you can debase a modern currency by printing, you could debase coinage by reminting. That tended to happen when wars needed to be financed. When the denarius was introduced in 211 BCE it was almost pure silver. Roman coinage eventually declined to the point just prior to 300 CE, about 500 years later, where the silver content of the denarius had fallen to 2%.

Bizarrely, from 768 CE, the time of Charlemagne, until the end of the 12th century, the denier was the only coin minted in Europe—just one coin for everything. The situation was similar in England, where the one coin was called the "penny." As you can probably imagine, a one-coin economy makes payments for both large and small things very awkward.

For example, a penny, a day's wages for an English laborer in 1250, bought 10 gallons of ale plus four one pound loaves of bread. So the penny was impractical for small quantities of anything. It was also inconvenient in large amounts. For example, the annual income of Canterbury Cathedral was 3.4 million pennies. That weighs in at about 3.3 tons.

The Galleons of Gold

Some people have a deep faith in gold as a currency, because the supply of new gold and the destruction of gold (through loss and industrial use) is almost the same and has been for a few centuries. Currently, about 188,000 tons of gold in pure ingots exists.

During the late 15th and early 16th century, there was a shortage of precious metals in Europe, but it soon gave way to a flood. In its conquest of the Incas, Spain took control of the Inca mines and also stole copious amounts of gold and silver, flooding its economy and the economies of Europe with money. The immediate consequence was inflation; the money supply increased and prices rose. Spain went on a spending spree and the combination of monetary instability and war expenditures, particularly the ill-fated Armada, led to the Spanish monarchy going bankrupt three times by the end of the 16th century.

Such precious metal instability is unlikely to happen again—but it is not impossible. There is an estimated 80 million tons of gold suspended in particle form in seawater. At present there is no technology which can mine it, but if there were, gold might cease to have a stable value.

Chinese Flying Money

Europe's 16th-century monetary inflation was mirrored by inflation in China. It was a coincidence, as there was no significant connection between the two economic spheres or their currencies. China had invented paper money, which it printed using wooden block technology and it had been using it since the Tang Dynasty (618-907 AD). It had also gradually developed credit mechanisms—merchants had begun to use promissory notes (credit notes) for long-distance trade.

During the Song Dynasty (960-1276) merchants were invited to deposit their coins with Government Treasury in exchange for Fey-thsian or flying money. A major motivation for this was that China was running out of coins - a money-supply problem that often plagued coin-based economies. Some merchants issued private drafts (bank notes in effect) that were backed by coins and salt, and later on by gold and silver. In 1024 the government took sole control of issuing these notes. Then during the Yuan Dynasty (1279-1367) paper money became the only legal tender.

Financial mismanagement began in the Ming Dynasty (1368-1644). New notes were introduced into circulation without withdrawing older notes. Monetary inflation followed, as naturally as night follows day. In 1380, one Chinese guan was worth 1000 copper coins, by 1535, its value had reduced to 0.28 copper coins.

European Paper Money

Europe didn't do much better with its first forays into paper money. Bank notes evolved from promissory notes as had happened in China. Promissory notes were issued to those who kept coinage or precious metals at a bank and could be redeemed only by the depositor.

A first short-lived attempt at paper currency was made in Sweden in 1661 by Stockholms Banco, a predecessor of the Bank of Sweden. Bank notes were at first linked directly with deposits. Then the manager of the bank unlinked the notes, allowing anyone who held the note to redeem the associated deposit. Next, the bank printed more notes than there were deposits. It went bankrupt after three years.

At about the same time, the goldsmith-bankers in London began producing bearer-notes, but they were far more conservative in issuing them than Stockholms Banco. They proved to be a prototype for a national currency.

In 1694 the English government established the Bank of England, in an effort to raise money for a war against France. It issued notes that "promised to pay the bearer on demand" a specific weight of sterling silver. Such notes were initially handwritten for a precise amount. Fixed denomination notes from £20 to £1,000 appeared by 1745. These were the first British bank notes. While a slow monetary evolution inched forward in London, more ambitious ideas were rapidly taking shape in Paris, under the guidance of the Scottish economist, John Law.

A Tale of Two Cities

John Law was an original economic thinker. He believed that money was only a means of exchange and that it did not constitute wealth in itself. He was the author of the scarcity theory of value—a model for the interaction between supply and demand. He wrote the Real Bills Doctrine, which considers the required relationship between bank assets and liabilities and which set the foundation for modern banking. He was of the opinion that money creation (i.e. money printing) would stimulate an economy - which is now known to be true in the short term. He regarded shares as "a superior form of money," because they paid dividends. He regarded paper money as preferable to metallic money, which it is, as we shall discuss later.

Law was not just an early economic thinker, he was also a successful gambler who understood probability and was proficient at mental calculation. It was his gambling skills that brought him to the attention of the French aristocracy and his economic ideas that got him appointed Controller General of Finances of France by the Duke of Orleans, who was regent for the young king, Louis XV. Thus began John Law's short and notable career in national economics, which ran from 1716 to 1720.

Louis XIV's War of the Spanish Succession, had left France in economic distress. Its national debt was crippling and the economy was stagnant. There was a shortage of precious metals and consequently a shortage of coins in circulation. Law's solution was to replace gold with paper credit and then gradually increase the supply of credit. This, he believed, would stimulate industry and then the national debt could be reduced by replacing it with valuable shares in prosperous economic ventures.

In May 1716 Law set up the Banque Générale Privée, a private bank that issued bank notes. Investors in the bank could deposit currency (Louis d'or gold coins) combined in a 1 to 3 ratio with defunct government bonds. The bank issued banknotes based on the gold coins. Law then tried to establish monopoly trading companies (like the English South Sea company) and succeeded by founding the Mississippi Company, a company that consolidated the trading companies of Louisiana under a single monopoly.

The share price of the Mississippi Company rose and then boomed. Ordinary French citizens began pouring their savings into it. Between May and December 1719 the market price shot up like a rocket, by a factor of twenty, and continued rising, eventually attaining the giddy height of 60 times its original price. Then it paused, and fell, and panic selling ensued. The public rushed to convert their banknotes to coin, causing a run on the Banque Générale Privée, which forced Law to close it for ten days and limit the transaction size when the bank reopened. Mississippi Company stock price continued to collapse, and the shares were eventually rendered worthless.

Meanwhile, because Law had been issuing banknotes to pay for Mississippi company stock, price inflation swept the economy. The money supply roughly doubled and by January 1720 price inflation had risen beyond 20% per month. Law was dismissed from his post at the end of 1720 and France returned to a coinage system.

Despite the ignominious failure of his French monetary innovations, John Law's theories still live on. His idea of a central bank that issued paper money was years ahead of its time. It would come to pass.

The Resistance to Paper Money

The evolution of banknotes proceeded sedately in London. In 1717, Sir Isaac Newton, who was master of the Royal Mint, switched to defining the pound's value in gold rather than silver. The Bank Charter Act of 1844 gave The Bank of England a monopoly for the issue of new banknotes, legally establishing it as a central bank and officially pegging the value of the British pound to gold.

Before that, what we think of as banknotes didn't exist anywhere; John Law's disastrous French experiment had proved to be a highly effective deterrent. The faith in coinage is even visible in the US constitution, where Article 1, Section 10 states:

"[No State shall] make any Thing but gold and silver Coin a Tender in Payment of Debts;"

That was a restriction on states rather than the US as a whole, preventing banknote experiments at the state level. After the founding of the US, the dollar was established as coinage with the implementation of a bimetallic standard with a 15:1 ratio of the value of silver to gold. Such a value ratio was clearly arbitrary but could be reviewed and changed. That occurred with the Coinage Act of 1834, when the ratio was changed to 16:1.

The US was far less enthusiastic about the idea of a central bank than the UK. In 1791 it established the First Bank of the United States, a private bank with a 20-year charter, but Congress never renewed the Charter in 1811. Five years passed, then in 1816, The Second Bank of the United States was formed and given a 20-year charter. When Andrew Jackson became president in 1828, he denounced the bank as an engine of corruption. Unable to get the bank dissolved, he refused to renew its charter. He also issued an executive order requiring all Federal land payments to be made in gold or silver. That policy may have been one of the causes of The Panic of 1937, a US financial crisis that lasted for over 6 years, but it was not the only cause.

From 1837 to 1863, a period of 26 years, the US had no central banks, just heavily regulated state-chartered banks which issued bank notes against specie (precious metal-based coinage). Then, with the National Banking Act of 1863, a system of national banks was established with the intent of creating a uniform national currency and paying for the Union's civil war costs. They issued "Greenback" dollars that were not backed by precious metals. In a series of legislative changes, which began in 1873, the US gradually moved to a gold standard with all bank notes backed by gold.

The Rise of the Gold Standard

The British had been operating a gold standard since 1844, and much of the British Empire naturally fell in line. In 1873 Germany adopted the gold standard and France followed suit.

Throughout the 19th-century, banknotes backed by gold gradually superseded gold and silver coinage. A gold standard emerged which had some useful characteristics. It prevented undisciplined and inflationary money-printing. It reduced the need for coins. It enforced a fixed exchange rate between currencies, facilitating global trade. Because currency was exchangeable for gold, gold naturally migrated to the more successful economies, increasing the money supply in accordance with the national balance of trade.

What's not to love?

Gold does not provide a perfect basis for a commodity-backed currency, but, historically, it has proved to be a very effective choice, despite various glitches. Currently the world's above-ground gold stock grows by 1 to 2% Annually. About half the gold in existence is used for jewelry, about 20% is owned by central banks, a further 20% by private investors and the remainder is employed industrially. The neat thing about gold is that it isn't destroyed by usage; almost all industrially used gold is recovered and recycled.

The glitches in gold-base currencies happen with discoveries of significant new deposits. These cause unavoidable inflation of all gold-backed currencies, no matter whether paper or coinage. The supply of gold increased, which meant that the money supply increased without a parallel increase in economic activity—so prices in the economy increased. It happened with Spain's influx of gold, but also from the California gold rush of 1849, the Australian gold rush of 1851 and the Witwatersrand gold rush of 1886. For example, after the California gold rush in the US, the money supply rose. In the period 1849-54 it rose by 109% resulting in price inflation of 32%. But as the gold rush faded, inflation returned to normal levels.

It would require a truly dramatic new gold find to increase the existing stock of gold significantly. There are currently, about 190,000 tonnes of mined gold, with an estimated 54,000 tonnes still below ground. Most of that 190,000 tonnes is more recent than you might imagine. Two-thirds of it has been mined since 1950. The supply situation is stable, with 2,500-3,000 tonnes added each year—an amount that is more likely to diminish than increase in the future.

The question is: why did the gold standard fail?

But it's the wrong question. The gold standard didn't fail; it worked. The intention with the gold standard was to prevent the debasement of a paper currency by making it exchangeable with something that couldn't be debased: gold. All the evidence suggests that it works, even allowing for the disruption imposed by the occasional major gold discovery.

The Destruction and Reconstruction of the Gold Standard

At the outbreak of World War I, the British Economist, John Maynard Keynes, predicted that it would end within a year, because none of the countries involved could afford more than a year of war. He was wrong.

Instead, the warring powers printed money to pay for their world war, sacrificing the gold standard in the process. And by the way, that wasn't a new idea; Abraham Lincoln paid for the civil war by issuing fiat currency, because there was no possibility of paying for it in real gold.

The first to drop the gold standard were the Germans, followed swiftly by all the other major combatants. The only country that remained on the gold standard throughout the war was the United States, partly because it remained neutral until close to the end. When it joined in, President Wilson banned gold export, effectively suspending the gold standard for foreign exchange. World War I devastated the European economies, and an attempt was made to restore global trade and return to a stable currency arrangement.

The Genoa Conference of 1922 proposed that central banks make a partial return to the lapsed gold standard, and the assembled powers agreed to a slightly different gold standard that "conserved" national gold stocks. Gold would remain in the vaults and central banks would additionally hold reserves in major currencies, which were exchangeable for gold coins—citizens were only able to exchange currencies for gold bars, a rule implemented to tamp down the personal demand for gold.

In the Great Depression, all the major currencies compromised this gold exchange standard. In 1931, in Europe and the US, speculators bought gold bars in exchange for banknotes, putting pressure on the local currency. The UK was the first to buckle, devaluing the pound against gold. In the US the Federal Reserve was forced to raise interest rates to protect the gold standard, which made the depression worse. Bank runs became common in early 1933, so people began to hoard gold coins, in preference to bank notes, depleting US gold reserves.

Franklin D. Roosevelt intervened dramatically, issuing Executive Order 6102, forbidding the hoarding of gold coin, gold bullion, and gold certificates within the continental United States. It banned the private ownership of gold, requiring all citizens to exchange their private gold holdings at the price of $20.67 per troy ounce, allowing for the retention of no more than $100 in gold.

Then on 30 January 1934, the US Congress passed the Gold Reserve Act, which nationalized all gold by ordering Federal Reserve banks to turn over their physical gold to the U.S. Treasury. In return, the banks received gold certificates to be used as reserves against deposits and Federal Reserve notes. The act authorized the president to devalue the gold dollar, which he quickly did. This bumped the value of the dollar from $20.67 to $35 per ounce, a devaluation of over 40%. The effect was to allow the Federal Reserve to print money in the hope of combatting rampant deflation.

After that the gold standard was effectively dead. France held on until 1936, but then it too threw in its hand.

Bretton Woods and the Reserve Currency Caper

In the absence of a gold standard, countries imposed exchange controls and fought trade wars to put a brake on adverse international money flows. It was a time of economic contraction and distress that paved the way to the Second World War. It is debatable as to whether the global economic environment caused the war, but had the gold standard remained in place, it would have curbed rather than facilitated military action.

Towards the close of the Second World War, in July 1944, a faux gold standard, called by some the "gold exchange standard" was established by the Bretton Woods Agreements. The idea was that most countries would fix their exchange rate against the US dollar and would be able to exchange their dollar holdings into gold at the official exchange rate of $35 per ounce whenever they so wished. Neither businesses nor individuals were allowed to do that, so convertibility was highly constrained.

This arrangement was very favorable to the US. It was the dominant economic power, and it had the leverage to craft the rules in its favor. The US dollar thus became the world's primary reserve currency—the currency that countries and large businesses needed to hold to conduct international trade.

They held dollars for the same reason that people and companies once held gold coins: as a common currency. The dollar was suitable because of its link to gold, and all countries could swap their dollar holdings for gold whenever they wished.

To fulfill the dollar's role as a reserve currency, the US needed to print an additional supply of dollars to circulate outside the US, and the countries and international businesses that needed them had to buy them. The US could print these dollars without fear of debasing the dollar within the US, because those dollars never came home. It could also run a large balance of payments deficit without damaging the short-term or even medium-term value of the dollar—and so it did. In all other countries, a persistent balance of payments deficit would depress the value of the currency.

A curious consequence of the reserve currency role of the US dollar is that the proportion of US dollars held abroad is not known—not even approximately. It is known that American households and businesses account for roughly 15% of the dollar stock and that a fairly large proportion of the stock lives invisibly in the US, hiding from the tax man or (in the case of ill-gotten gains) from the long arm of the law. The remainder are circulating abroad, with estimates varying between 33% and 66% of the total.

It is possible, one day, that those chickens will come home to roost.

Abandoning Gold in Favor of Fiat

Starting in 1959 and continuing for over a decade, France decided to continuously exchange its dollar reserves for gold at the official rate. In the latter part of the 1960s, the US was spending heavily on the Vietnam War and was running persistent balance of payments deficits. The US was obliged to cut the deficits or cut the link to gold. President Nixon chose the latter—on August 15, 1971, the US abandoned the convertibility of the US dollar to gold.

As a direct consequence, not only the US dollar, but all other currencies in the world, became fiat currencies. Just for show, the dollar was immediately devalued by resetting the gold price to $38 per ounce and, two years later, in October 1973, to $42.22 per ounce but convertibility never returned. So those prices were meaningless, as soon became clear in 1974 when the Chicago Mercantile Exchange (CME) began trading a gold futures contract.

The graph on the next page tells the story of the gold standard. As we noted, the gold standard works if the intention is to hold the money supply roughly constant. The vertical axis on the graph measures price inflation. Price inflation is usually caused by increasing the money supply. There can be other causes such as poor harvests, but historically it is usually money printing that causes price inflation.

The chart presents strong evidence that whenever there is a need to

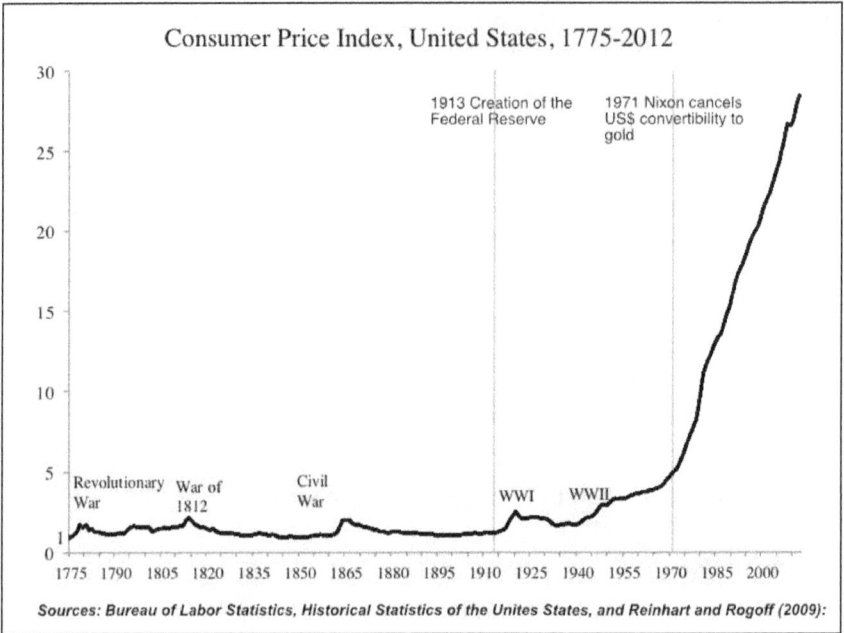

Consumer Price Index, United States, 1775-2012

1913 Creation of the Federal Reserve

1971 Nixon cancels US$ convertibility to gold

Revolutionary War

War of 1812

Civil War

WWI

WWII

Sources: Bureau of Labor Statistics, Historical Statistics of the Unites States, and Reinhart and Rogoff (2009):

Figure 3. The Impact of the Gold Standard and its Abandonment

finance a war, governments start to print money. Prices rise, and when the war is over, the money printing stops and the prices fall back again. WWI and WWII are visible in the graph, as is the price deflation of the 1930s depression. From the beginning of the Bretton Woods agreement in 1944, prices began an upward trajectory, which headed for the stratosphere after the dollar's link to gold was cut. We could draw a similar graph for other currencies, some of which have printed money far more aggressive than the US Federal Reserve. As things stand, it is possible for a country to maintain currency parity by not printing money any faster than the US is doing.

Chapter Summary

Chapter 4 focuses entirely on the history and nature of money.

- For a brief few years following the end of the World War II, cigarettes served as a currency in Germany. This occurred simply because there was no acceptable currency available in this period and hence cigarettes self-selected as a currency that was well distributed and had a known value.

- There is an important difference between **bearer money,** money that the bearer owns by virtue of possession, and **account money**, money that is held on their behalf in a specific "account." The historic record suggests that both types of money date back to prehistoric times.

- Coinage has the virtue that the coins can be a store of value and hence do not depend on the "faith" of the holder. They can thus also have value outside the regime that issued the coins. However, their weight poses problems and so does coin fraud.

- Paper money began in China and, after a false start in France, was solidified in the UK through a link initially to silver and later to gold.

- The gold standard was eventually adopted by the major economies, but was abandoned by participants in WWI so they could finance their war.

- At the close of WWII the Bretton Woods agreement tied all currencies to the dollar, with the dollar being tied to gold. This arrangement persisted until President Nixon cut the link between gold and the dollar. So now no currencies are backed by the intrinsic value that gold represents.

- The gold standard did not fail, it was abandoned.

Chapter 5

Let There Be Money

Today is yesterday's pupil.

~ *Benjamin Franklin*

—⁂—

Although they were prolific thinkers across the spectrum of philosophy and science, the Ancient Greeks did little thinking about economics. They scorned the idea of trade. They viewed it as the domain of profiteers, who sought to buy goods cheap and sell them dear. That may seem a little eccentric, until you realize that the Greeks conceived of their city states as governed by a civilized and cultured leisure class, served by slaves who attended to their material concerns. In practice, they were not industrious economies.

Nevertheless, Aristotle did eventually stoop to thinking about money. He concluded it should have the following characteristics:

- **Durability:** It must not wear out. In fact, everything physical that has been used as a currency does wear out eventually, but not wearing out quickly is a good thing.

- **Portability** (transportability): This is practical and is an obvious problem with coinage, where the large amounts of coins are difficult to transport, but not paper money

- **Divisibility:** This is about the denomination of a currency. Ideally you want to trade multiples and fractions of the basic unit.

- **Fungibility:** This is a fairly obvious requirement—fungibility means interchangeability with identical items. If I give you a coin, you only need to pay me back with a coin or coins of the same value, not the exact same coin.

- **Intrinsic Value:** Money should have value of itself, such as coins being minted from precious metals.

Aristotle's thoughts on money provide a historic context—he seems to have been the first individual to comment on money. However his thoughts are limited to coinage. He makes no mention of ***account money*** and his requirement for intrinsic value is impractical in modern times, although it made sense in his day.

The idea of paper money was beyond the orbit of his imagination. The same can be said of modern economists in respect of digital money and the blockchain—few are aware of the technology and its possibilities. But before we bite on that, let's sink our teeth into the elephant in the room. His name is debasement.

The Debasement of Currencies.

Currency debasement is an unsolved problem. On one hand, money is under perpetual attack from counterfeiters. If they succeed, they not only reap personal rewards, they also debase the currency. Counterfeiting is regarded as a serious crime and the penalties can be severe, it is not only theft, it debases the currency.

And yet we have governments, whom, as history demonstrates, see debasement as an alternative to taxation when they are strapped for cash. Debasement of paper money is achieved by money printing, usually indirectly by issuing government bonds, buying back bonds already issued, or paying interest on those bonds. Debasing a currency this way takes time and, in small doses, may not even be noticed; taxes take effect at once. They are visible and unpopular.

In practice, there is little difference between the Roman Empire debasing its silver coinage and Britain or Germany abandoning the gold standard. The main intellectual attraction of the gold standard is that it makes currency debasement either visible or impossible, depending on the governance rules.

Three Types of Money

The solution to preventing currency debasement is put it in the hands of an authority that will not debase it. To discuss the possibilities, we need to discuss the nature and technology of money, including cryptocurrencies, in depth. In doing this, we have had to create a new classification of money which takes account of the nature of cryptocurrency. We deliberately avoided the M0, M1, M2 classifications of money supply that is commonly used by governments and economists. We will explain why in due course.

From the perspective we wish to consider, there are three types of money:

- **Bearer money.** This is money that has no knowledge of who owns it. Whoever carries it is considered its owner.

- **Account money.** This is a store of money held somewhere on the owner's behalf. If its storage location is fully secure it cannot be stolen, unless it is removed from its store—at which point it becomes *bearer money*.

- **Fractional reserve money.** Can be thought of as "invented" money; the product of fractional reserve banking.

In almost every financial environment these types of money co-exist, each carrying a different kind of risk. *Bearer money* offers convenience, but with the risk of theft. In the days when there was only coinage, there was a limit to how much could be conveniently carried. *Account money* is less convenient, but it is more secure and there is no real limit as to amount. *Fractional reserve money* can be considered *account money* for most practical purposes—it is secure and not limited.

The value of these three types of money is almost identical, in the sense that a dollar held in a bank account is worth exactly as much as a dollar held in your hand. There is a potential difference in value, since the account service may levy transaction costs on transactions or, alternatively, may pay interest for money held on account.

The Nature of Bearer Money

Currency debasement manifests in coinage through techniques such as clipping the coin edges or reforging the coins with less precious metal content. In the case of paper money, debasement occurs by forgery or money printing. When a currency is debased the money supply increases. If the amount of economic activity within the economy remains the same then price rises soon follow.

If the level of debasement is small, the impact of rising prices will be slow. You can think of it like this: If the government increases taxes on everyone by 3%, most likely citizens will be annoyed and may reflect their disapproval at the ballot box. Instead the government can simply print 3% more money and can harvest the same amount of money for itself without attracting much attention. Conveniently, the impact in rising prices may not become apparent for 12 to 18 months. And if the economy is growing by 3% then even better, because the increase in the money supply will not necessarily be visible at all.

Bearer money can and does move from place to place, and can even travel out of the country. This is a particular problem with coins whose value in use is close to the value of their precious metal content. The coins may simply exit the country. In effect, a balance of trade problem suddenly becomes a coin shortage problem.

Bearer money is also the criminal's natural choice. Its ownership is never known and it can pass from one hand to another invisibly. From our perspective, money laundering is the act of disguising the transfer of criminally earned *bearer money* into *account money*.

Paper notes and coins can be lost or destroyed, and paper notes wear out quite quickly. For dollar notes, the lifetimes are: $1—about 5.8 years, $5—about 5.5 years, $10—about 4.5 years, $20—about 7.9 years, $50—about 8.5 years and $100—about 5.5 years.

The constant printing activity costs about 5 cents (for $1 and $2 bills), about 10 cents (for $5, $10, $20 and $50 bills) and 12.3 cents for $100 bills. It doesn't sound particularly expensive, but for every $1 bill it costs ¢1 to keep it in circulation for a year, and as there are 11.7 billion such notes, we're looking at over $110 million per annum just for $1 bills.

Paper notes are relatively cheap to make; coins are not. In the US, the price varies slightly with the market price of copper and nickel. In 2014 it cost $1.66 to make a dollar's worth of pennies, $1.62 to make a dollar's worth of nickels, 40 cents for a dollar's worth of dimes, 36 cents to make a dollar's worth of quarters, and 10 cents to make a dollar coin. The penny and nickel have for many years cost more than their face value to make.

Ignoring half-dollar coins, the annual US coin production is roughly 16 billion coins, with about 56% pennies, 10% nickels, 19% dimes, 15% quarters and 0.06% dollars for a value of $1.7 billion, at a cost of about $134 million. Coins can last up to thirty years, so it would make great sense for the dollar coin to replace the dollar bill, but the public prefers the note.

The cost then of circulating paper money and coins is significant, and governments would be happy to do away with it. In the US this is made impossible by the fact that about 7.7% of US citizens are unbanked and thus cannot use the digital payment services that banks provide. Cash has a dominant role in small-value transactions and is the natural alternative when other payment options are not available.

The Nature of Account Money

At a basic level, you can think of *account money* as a deposit of *bearer money* held in a safe store for you. It probably started out that way, but it never stayed that way. The guardians of *account money,* which was often gold coins, evolved into banks. They lent your deposits out to others, who were able to use the money in the form of an interest-bearing loan. To incentivize you to store your money with them, they would pay you interest too, at a slightly lower rate. The difference between the two interest rates covered their costs and allowed for some profit.

That's how fractional reserve banking evolved, with banks only needing to hold a proportion of the money they loaned out as a reserve against the regular flow of people drawing down their deposits. To understand *account money* and fractional reserve banking, think in terms of two kinds of *account money*, typified by a checking account and a savings account.

Typically, users of a checking account put some or even all of their earnings into the account regularly. They use it to pay bills and meet daily expenses. They can write checks on the account or can charge the account directly with a debit card. The bank makes charges against the account for the service it provides. The cost to individuals of writing a check varies, but is usually low. The cost to business users averages out at about $6.00 per check. Debit card costs vary, usually on the basis of a fixed cost plus a percentage of the amount—for example $0.15 plus 0.80% of the amount.

So with *bearer money* the cost per transaction is zero, but with *account money* there is a per transaction cost. Thus it is logical to use *bearer money* for small transactions and checks and debit cards for large ones. In general that's how people behave (in the US).

The Nature of Fractional Reserve Money

The savings *account money* acts as a reserve for loans the bank makes. In effect the bank acts as an intermediary between borrowers and savers, and it loans far more money than the deposits it takes from savers. As a consequence, the bank can experience a bank run when depositors try to withdraw more funds than can be covered by the reserves that are held by the bank. Governments regulate and monitor the banks to try to ensure that they keep adequate reserves. If a run occurs they will normally act as lender of last resort to prevent the bank from collapsing.

Banks tend to hold reserves in a ratio mandated by the government, for example 1:10, meaning that one dollar is held in reserve for every 10 loaned out. What this means in practice is that by making loans the bank is increasing the money supply. In general, fractional reserve banking allows the money supply to grow well beyond the amount of the underlying money originally created by the central bank. We refer to this and can think of this as *fractional reserve money*. Eventually it will become real money (i.e. *bearer money* or *account money*) or else it will disappear when the borrower proves unable to pay.

The typical cycle is as follows: for a number of years loans are made, money is invested and the economy grows. The amount of *fractional reserve money* increases and some of it changes to become bearer or *account money*, as the result of profitable activity. Eventually there is a recession and some of the *fractional reserve money* simply vanishes, reducing the money supply. And then the whole cycle begins again.

Money Supply: M0, M1, M2, M3

There is now detailed consensus among economists, how the money supply ought to be measured. There are different measures: M0, MB, M1, M2, MZM, M3, M4-, M4, and L. How they are defined and measured even varies from country to country. For the sake of simplicity, here we discuss the US versions of M0, M1, M2, and M3, explaining what they mean in terms of *bearer money*, *account money* and *fractional reserve money*.

- **M0:** This is *bearer money* and thus it provides the thinnest measure of all money within an economy. Figures for this measure for the US are misleading, because (in the estimation of the Fed) two thirds of that M0 money is not circulating within the US. The ratio between M0 and M1 money supplies is fairly meaningless anyway, as it depends on other variable factors too, such as the preference of people for coins over notes, the demand for coins for slot machines and so on.

- **M1:** This measure includes checking accounts, and thus includes the "liquid" money that regularly circulates in the short term and keeps the economy ticking over. M1 is also taken to include all of M0 and it also includes demand deposits, ATM money, and travelers' checks. If we exclude the *bearer money* (M0), this is the money we chose to call *account money* in this book.

- **M2:** This measure of money includes all of M0 and M1. In addition, it comprises savings deposits and any equivalent type of deposit where the money is generally not frequently withdrawn, and from which what we have called *fractional reserve money* can arise.

- **M3:** This measure includes M0, M1 and M2, but also includes large and long terms deposits. Large and long-term deposits can also be a source of *fractional reserve money*.

There are two further points worth mentioning here.

1. We note that fractional reserve banking can and does make use of the rolling balance of money in checking accounts, in addition to the money in savings accounts, as deposits against which it can make loans. Thus, to some degree, all the non-M0 components of M1, M2, and M3 include *fractional reserve money*.

2. We have not mentioned central bank activity. We can consider commercial bank money and the *fractional reserve money* it creates as distinct from central bank money. Central banks have two tools they rely on to implement monetary policy. One is the central bank interest rate (the Fed funds rate in the US). When the central bank wants to print money it lowers its interest rate.

Banks are generally required to hold their reserves in short-term government funds. When the interest rate is lowered, the banks pay less for their reserve and thus they can lend more at lower rates. Similarly, if the interest rate is raised they need to lend less and increase rates. Thus the central bank can stimulate the economy or cause it to cool. The central bank can also buy government bonds and other securities from banks and replace them with credit, in effect, inflating the money supply (because it simply invents new money that didn't previously exist). This is simply a way of printing money, which the Federal Reserve euphemistically calls "quantitative easing."

It is also worth noting that central banks hold reserves, usually of precious metals (gold and silver), as well as amounts of foreign currency, usually in government bonds. These reserves are increased or decreased at various times, generally in response to any rapid change in the exchange rate against other major currencies. They are, in effect, a buffer, but not an inexhaustible one.

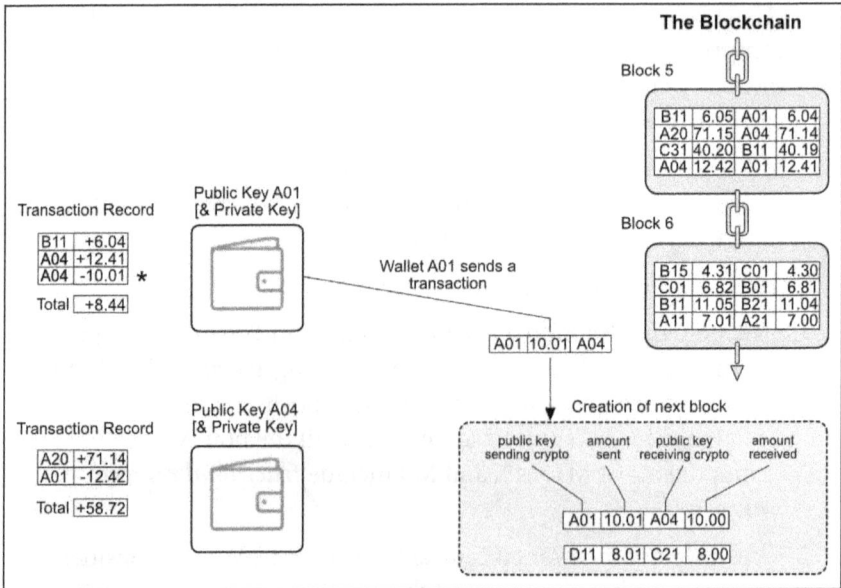

Figure 4. A Simple View of Wallets and the Blockchain

The Cryptocurrencies

Cryptocurrencies are **account money**. They store money in accounts and allow transactions on those accounts. We illustrate this in *Figure 4* above. It shows two wallets and a blockchain. The most important thing to understand is that every account that exists, and all the transactions for all those accounts, are stored on the blockchain. The blockchain is a ledger and every account that exists which holds cryptocurrency has all its details on that ledger.

We illustrate this in the diagram by showing blockchain transactions. In our example, each block in the chain stores just four transactions, and each transaction consists of a movement of cryptocurrency from one account to another. So, for example, in Block 5 the first transaction records that account B11 has transferred 6.05 units of crypto to account A01. Note that account A01 receives only 6.04 units. That's because there is a transaction charge for each transaction. This is money that is retained by the "miner." We use the term "miner" to denote the server that completes that transaction block. Completing blocks, no matter how it is achieved, can be thought of as "mining." It has nothing to do with real mining. It is a metaphor for a rewarding activity.

70

These account names, B11 and A01, are in reality public keys from a PKI public/private key pair. In practice, both accounts' names would be meaningless and long, e.g. KjlyJuNnN7TinNP99InmHTcZR8yTHT. Illustrating it with a short name makes it seem less complicated.

Figure 4 also illustrates two wallets, A01 and A04, with A01 sending a transaction to the blockchain to transfer 10.01 crypto units to A04. It is important to understand that although these software programs are called "wallets," they never contain any crypto units. All of the crypto units are and will forever remain on the blockchain. The wallets make it possible for the owner of some cryptocurrency to send it to or receive it from other owners of cryptocurrency.

It works like this:

The wallets have two keys, a public key, and a private key. When your wallet sends a transaction, it is encrypted with the private key that only your wallet knows. The computer servers that store the blockchain can decrypt and read the transaction using the public key. They know it is genuine precisely because it could only have been encrypted with the private key. This means that only your wallet can create a transaction on your crypto account on the blockchain.

The wallets are created by their owners, from a "seed," a code which consists of a unique string of words. When the words are fed into the wallet software, a wallet is created that has a specific public and private key. If the wallet was somehow lost or deleted it can be recreated from the seed. All the information that the wallet stores, aside from the private key, is also contained on the blockchain. So if the wallet is regenerated it simply reads blockchain records to get the information it needs to restore its state. These would be all the transactions on the blockchain which refer to the wallet's public key.

The wallet contains a list of all the transactions recorded against its public key, and it can thus calculate how many crypto units are stored under that public key. When a wallet owner wishes to send an amount of crypto units to someone (to some other wallet) they enter the transaction amount into the wallet. The transaction is then encrypted and sent to the blockchain network. *Figure 4* illustrates a block being created from several transactions. Wallet A01 is sending a transaction of 10.01 units to wallet A04. This is then added to the block that is being created. When, in this example, four transactions have arrived, a new block can be created and added to the end of the chain to create Block 7.

If you look again at the diagram you will notice that the transaction made by wallet A01 is stored in that wallet with an asterisk which indicates that this is a transaction that has been sent but has not yet been confirmed. Similarly that transaction is not yet shown in wallet A04, because it has not yet been confirmed. It will be confirmed when the new block has been created, and then wallet A04 will be able to register the transaction.

A blockchain network consists of many servers, each of which stores a full copy of the blockchain and is capable of creating new blocks. So in practice a wallet will send the transaction it creates to every server and each will work on creating a new block. The metaphor of "mining" for creating blocks was used with Bitcoin, because in that cryptocurrency, when a new block is completed, the server which completes the block is rewarded with newly issued Bitcoins.

Not all cryptocurrencies work in exactly that way, as we shall discuss later, but with every blockchain there is always a population of servers that can create new blocks, and only one of them will succeed and be rewarded for doing so.

We have said enough about cryptocurrencies here to allow us to make the following assertions:

- A cryptocurrency is *account money*.
- It can never be *bearer money* because, by its very nature, it knows who owns it (the private key owns it).
- Neither can it ever be *fractional reserve money,* as there is no possibility that new units of the cryptocurrency can be created from existing ones in the way that banks create new units of currency when they carry out fractional reserve banking.

Chapter Summary

Chapter 5 deals with the characteristics of money that make it practical.

- Currency debasement, both in respect of the fiat currencies of the modern world and the coinage of the ancient world, is an unsolved problem. Counterfeiting can be minimized but there is nothing to stop any government from debasing its currency. The problem is clearly in the governance of the currency.

- In general, there are three distinct kinds of money:

  > **Bearer money.** This is money that has no knowledge of who owns it. Whoever carries it is considered its owner.

  > **Account money.** This is a store of money held somewhere on the owner's behalf.

  > **Fractional reserve money.** We can think of as "invented" money, the product of fractional reserve banking.

- There are specific measures of the fiat money supply in an economy. Exactly how money supply is measured varies between nations, but in general the measures: M0, M1, M2, and M3 are used. These measures do not distinguish between the types of money that we have described. The virtue of these classifications is that they deal with the "availability" of money (the speed at which it can be transacted).

- Cryptocurrencies are account money. A cryptocurrency wallet is equivalent to a personally controlled bank account.

- By their nature, cryptocurrencies cannot be *bearer money* or *fractional reserve money*. The cryptocurrency enthusiasts who believe that cryptocurrencies will replace fiat currencies are dreaming.

The "Common Sense" of Cryptocurrency

Chapter 6

Crypto and Fiat Compared

Happiness is not an ideal of reason, but of imagination.

~ *Immanuel Kant*

———⊗⊗⊗———

In summary, cryptocurrencies will be complementary to fiat currencies, replacing them in some contexts but unable to do so in others. With that understanding, we can now explore the characteristics of fiat and crypto currencies from that perspective, and compare the virtues and defects of each.

We can define the characteristics of a currency with the following seven practical assertions:

1. A currency is a medium of exchange

A currency must be easily exchangeable for goods and services everywhere within its native economy. It does not necessarily have to have intrinsic value, but its users must accept it as having "value by use." When there is no medium of exchange, the citizenry will usually invent one, as happened in Germany. In some countries there will be legislation that declares the national currency to be legal tender, so shopkeepers and merchants are obliged to accept it.

Note that a cryptocurrency fails as a medium of exchange in situations where digital interaction is impossible (as do credit cards and debit cards). Paper money and cryptocurrency both fail in situations where one party to a transaction insists on a currency having intrinsic value (like gold.) What serves as a medium of exchange is thus contextual to some degree.

2. A currency must be portable

Portability can be a problem with coinage. When a large amount of money is needed, its weight imposes limitations on how to gather it and move it—which in turn can impose a significant cost and time limitation on executing a transaction. Paper money solved that problem and reduced the need for coinage significantly. It made it unnecessary to use gold in coinage, and it also reduced the amount of silver required.

Bearer money is, however, steal-able and losable. *Account money* in the form of debit cards (or even checks) is more portable for most purposes than paper money, and with the advent of payment software on mobile phones, it is as portable as the phone. Its primary limitation is that it requires the other party to the transaction to have compatible technology. Cryptocurrency wallets have the same limitation if you wish to send money directly, wallet to wallet. However, the proliferation of debit cards for crypto wallets will eventually mean that cryptocurrency has the same technology options as fiat currency.

From the perspective of portability, *fractional reserve money* and *account money* are identical.

3. A currency must be inexpensive to create, maintain and use

We have already brushed up against the fact that the costs of minting and managing *bearer money* (such as the dollar) can be considerable. No matter what the nature of the currency, metallic, paper or digital, there is the cost of creating the currency, the cost of ownership (or storage) and the cost of exchanging it for goods or services. How those costs manifest within a single transaction can vary. Nevertheless they are costs that someone has to pay. For example, debit card and credit card costs are usually charged directly to the retailer by the card company or bank. The retailer naturally compensates for those costs by raising prices to the consumer.

In most countries, including the US, cash transactions make up as much as 30% of all transactions—the figure can be much higher in some areas of retail. For the retailer, accepting cash is a necessity, but cash handling costs are high. Aside from "check out" staff and a supervisor to manage and monitor all cash activities, there may be a need for a cash room or vault for cash kept on premises. Cash management activities can be extensive, including: preparing floats each day, handling change requests, skims, and cash pulls, reconciliation, preparing deposits, and delivering the money to the bank. In medium to large organizations, there are also cash audits, discrepancy investigations and shrink investigations.

You can add the cost of secure transport to the bank for cash, and supplies of coin wrappers, bill straps, deposit envelopes and bags. You can also add the cost of accepting fraudulent bills or the cost of counterfeit detectors—take your choice, plus whatever the bank charges you for its cash service. As is clear from this inventory of possible costs, **bearer money** involves a significant cash management cost.

Retail banking has never been a particularly competitive sector, so there has never been much pressure to push down the "banking cost" of transactions, whether in cash, by check or by card. This, more than anything else, has thrust cryptocurrency to the fore, and it is this that is now causing concern in retail banking.

The fact is that cryptocurrency transactions usually have far lower transaction costs than comparable fiat currency transactions. In general, cryptocurrency activity is fully automated and there are no staffing costs involved. The blockchain provides the public ledger of all activity and also all of the accounts. The cryptocurrency wallet has a cost of acquisition of zero and it runs on the customers' computer or mobile phone at a minimal marginal cost.

It is not much different from a traditional bank account, aside from its very low costs.

4. A currency must not be easy to forge or steal

A currency that can be forged is flawed. National mints wage a perpetual war against counterfeiters, employing increasingly sophisticated coin and print technology to outwit the forgers. It's a technology race. In the days when coins were the only form of currency, forgery was a recurrent problem. Thousands of years later it is still a problem.

For digital currency, whether fiat or crypto, hacking is almost an equivalent problem to forgery. Banks have a good but not perfect record of avoiding mass digital heists. One of the worst was perpetrated on CitiGroup in 2011, with hackers gaining entry by imitating URL changes that corresponded to the entry of valid usernames and passwords. They got access to the accounts of about 200,000 people and made off with about $2.7 million, along with a haul of personal data.

Such heists are rare compared to cryptocurrency heists, which are more frequent with greater rewards. The daddy of them all (so far) was the filching of $473 million from the Mt Gox crypto exchange in Tokyo in 2014. At the time this haul constituted 7% of the total Bitcoin supply. It was achieved by what is called a "transaction malleability" attack. Simply explained, the hackers changed the details of Bitcoin transactions (to their own advantage) before they were included in blocks. Once included, the false transactions were immutable and the associated Bitcoin was immutably stolen.

Perhaps the second most famous crypto hack was the "The DAO Hack" which rocked the Ethereum world. The hack caused the Ethereum dev team to fork the Ethereum blockchain—in effect creating a new Ethereum currency. Technically, the primary difference between Ethereum and Bitcoin is that the Ethereum ecosystem enables the definition and execution of "smart contracts." Smart contracts are software that is directly attached to a blockchain. "The DAO Hack" compromised the Decentralized Autonomous Organization (DAO) a complex smart contract.

The business idea implemented by the DAO was to create a decentralized "venture capital fund" that could fund all future Distributed Applications (dApps) that would run in the Ethereum Ecosystem. The DAO's ICO was immensely popular, raising $150 million worth of Ether. There was, however, a flaw in the DAO contract code, which an enterprising hacker noticed. He quickly exploited the error, siphoning off one-third of the DAO's funds in a few hours.

This hack caused a practical and philosophical crisis in the Ethereum community. The point was this: if a smart contract is "the law" and it happens to have a flaw in it, then there is nothing illegal about exploiting the flaw—"caveat emptor" as the Romans used to say. Nevertheless, $50 million had found a new owner and the Ethereum community was not inclined to enable the perpetrator of this unexpected funds transfer to profit from it. The decision was taken to hard fork the Ethereum blockchain into the Ethereum Classic blockchain (ETC), which included the missing $50 million, and a pristine Ethereum blockchain (ETH), which did not. Curiously this did not prevent the "hacker" from capitalizing on his (or her) smart contract attack. It simply created a version of the Ethereum blockchain where the $150 million remained intact.

There have been other massive crypto hacks. The Hong Kong-based cryptocurrency exchange, Bitfinex lost $72 million in Bitcoin in August 2016. In December 2017, the Nicehash cloud mining service was hit for $60,000,000 and in January 2018 the Coincheck exchange lost $534,800,000 of the cryptocurrency NEM. These are not the only examples, just the big ticket ones. Significant hacks in the crypto world happen every few months, usually to less prominent and utterly unregulated crypto exchanges.

We are describing digital theft, not forgery. Nevertheless, these successful thefts have a negative impact on the faith anyone might have in a cryptocurrency. The money held in traditional bank accounts is, by any reasonable measure, safer from theft than money held on crypto exchanges. This is a natural consequence of poor security standards, which is usually attributed to the lack of regulation of crypto exchanges. The situation is ironic because blockchain technology itself has proved to be as secure as it gets, and then some—with no record of it ever being compromised.

However, with cryptocurrency, you, the wallet holder, are the banker, and you bear the cost of any successful theft of crypto from your wallet, or from any place you sent your crypto—such as an account on a crypto exchange. There are very secure crypto wallets (hardware wallets, such as the Trezor and the Ledger) which minimize the risk of theft almost to zero. At the time of writing, there are no crypto exchanges that integrate directly with such technology, but we expect such developments to occur. The Wild West era of the crypto world needs to come to an end, and it will.

5. A currency must be easily divisible

Small coins are the smallest usable units of any national currency. Because currencies tend to inflate, eventually the smallest coins are withdrawn from circulation as, for example, the English farthing and half-penny were withdrawn. The various coin designations are usually chosen to make it easy to pay any small amount using just a few coins. Fiat currency is usually accounted with two decimal places although some have none and a handful have three, with the final decimal corresponding to the smallest coin.

Cryptocurrency is different. Bitcoin was launched with eight decimal places and most other cryptocurrencies have imitated that level of precision. There are several advantages to that. First, as has happened with Bitcoin's price rising beyond $10,000, eight decimal places still provided the ability to represent amounts as small as one hundredth of a cent. Because the currency has no physical existence outside the digital world, it would even be possible, by amending the blockchain software, to add further decimal places if needed.

With a digital currency, the limit on the viability of transactions is proportional to the cost of processing the transactions, which can be far lower than processing fiat currency transactions. It is thus possible to implement "microtransactions," much smaller transactions than those that occur with fiat currencies. If computer technology continues to reduce in cost and increase in speed, as it has done for decades, then the cost of processing cryptocurrency transactions will continue to fall and cryptocurrencies will be even more divisible than they are.

6. A currency must act as a metric of value

Goods and services seek a common metric by which the value of one can be compared to that of another: "This bottle of wine costs $10; that one from the better vineyard commands a price of $22."

We have an inner psychic mechanism that assigns value to things based on the currency we know and our shopping experience. It is very difficult to displace that metric, just as it is difficult for those who were raised to gauge temperature in Fahrenheit to think in Centigrade. You quickly discover how difficult it is to switch your sense of value when you move to live in a different country, where relative prices are different, taxation is different and supply and demand are balanced in a different way.

Ideally, the value of a currency never wavers and everything else of value within the economy does, or can waver.

There is a bad corn harvest, demand exceeds supply and so the price of corn rises to $4.50 per bushel. Next year the harvest is good and the price falls to $3.40. If the currency behaved perfectly we would never entertain the possibility that the value of the currency itself had shifted in that year. We would know that last year's price and this year's price reflect only the supply and demand fluctuations of the corn market.

No currencies have ever been as perfect as that. The historical records suggest that the British Pound exhibited great stability throughout the 19th century, losing it when it abandoned the gold standard. In recent times, the Swiss Franc has offered a haven of stability, in comparison to other currencies, but its value has still fluctuated significantly year-over-year.

The challenge a currency that is stable within its own economy faces, is that its stability can be undermined by other currencies in the global economy. At a national level, the only way to prevent such disruption is to implement import controls of some kind, but this can have unwelcome

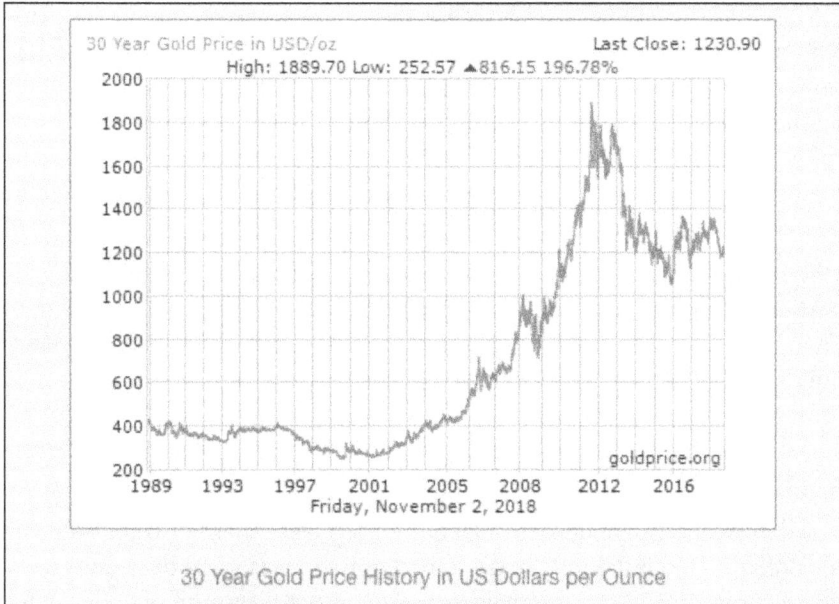

30 Year Gold Price in USD/oz — Last Close: 1230.90
High: 1889.70 Low: 252.57 ▲816.15 196.78%

goldprice.org

1989 1993 1997 2001 2005 2008 2012 2016
Friday, November 2, 2018

30 Year Gold Price History in US Dollars per Ounce

Figure 5. The "Unstable" Price of Gold

consequences, such as provoking a trade war. The only alternative to this is to debase the national currency slightly, but less so than comparable currencies. You are obliged to conform unless your economy is entirely self-sufficient.

The same challenge faces any cryptocurrency that wishes to be stable. Even if its price is not subject to the speculative activity on cryptocurrency exchanges, it can only be judged stable in respect of other currencies that are regarded as stable. Consider gold, for example. If you think of it as a currency then, in terms of money supply, it provides a shining example to every national currency on the planet—its supply has inflated by only about 2% for many decades. Compared to any currency in respect of "money supply" it is a paragon of stability.

And yet gold appears horribly volatile when compared to the dollar as illustrated in *Figure 5* above. This is because people treat the dollar as a metric of value. They project the volatility of the dollar onto the behavior of the gold price and believe that the value of gold is fluctuating when in reality it is not. This belief in a specific metric of value, which millions of people hold in common, presents a very significant challenge to the adoption of cryptocurrencies.

THE "COMMON SENSE" OF CRYPTOCURRENCY

Most cryptocurrencies are hopeless metrics of value, fluctuating in price wildly, sometimes by big percentages in a few hours. Their fluctuations are caused almost entirely by market speculation. In time, when the crypto market becomes less exuberant, some could become better metrics of value. However, as the commonly accepted metric of value (the dollar in the US) is not a paragon of money supply virtue, it is not clear how it could be replaced. It's a complex situation.

There are "stable-coins," cryptocurrencies whose value is pinned to a particular fiat currency, with the idea that they can be true metrics of value. The problem with such cryptocurrencies is that if the fiat currency fluctuates in value then so does the so-called stable-coin. As such, there is little point in such a stable-coin unless it has a different tax status to the currency itself, which currently seems to be the case in the US. However, that tax status could easily be changed.

Tether was the first stable-coin. In theory it was tethered to the value of the dollar on a one-for-one basis, although there was no publicly auditable data to prove the coin was indeed linked to an equal amount of dollars held on account. For a while Tether was the only such cryptocurrency. But in 2018 a glut of US dollar based ones emerged: Basis, Dai, Havven, TrueUSD and two high-profile ones, GUSD from the Gemini Exchange that is run by the Winklevoss twins and USDC from Coinbase and Circle. There is also EURS, a Euro-based stable-coin, LBXPeg, a GBP-based stable-coin, and a whole host of gold-based stable-coins too numerous to list.

Occasionally, the idea of a metric of value and a store of value are confused with each other. It is important to differentiate between these two characteristics. If a currency acts as a true metric of value, by maintaining a stable value, it becomes a non-appreciating "store of value." A currency should not be designed to appreciate. Financial instruments that are designed for appreciation and present minimal risk (such as short term government bonds) are better stores of value than a currency can ever be.

Central banks maintain currency reserves in gold, silver and other trading currencies in order to stabilize the national currency. This is done to improve the stability of the currency when there are trade imbalances. A trade deficit will naturally depress the value of a currency and a trade surplus will increase it. The gold and currency reserves are used to dampen those effects. In times of surplus the currency will be used to buy gold and other major currencies and in times of deficit those reserves will be reduced. It is likely that, as cryptocurrencies mature, their stability will be managed using similar kinds of reserve mechanisms.

7. A currency requires governance

Governance is a critically important factor that is usually omitted when the characteristics of a currency are discussed. History provides numerous examples of government deliberately debasing the currency they claim to be managing. We already mentioned the demise of the gold standard in the spending spree that went by the name of the First World War. The Roman Empire, in its declining years, ceased its conquests and debased its currency. England's Henry VIII, in an event referred to as The Great Debasement, sacrificed the English pound to finance his lavish lifestyle and wars with France and Scotland. History teaches us again and again that it is governments that destroy healthy currency systems.

This cannot, or at least should not, be the case with cryptocurrency. Self-governance is one of the fundamental foundations of a genuine cryptocurrency. This is so much the case that there are frequent debates and discussions in cryptocurrency forums over the best technical mechanisms to implement and enforce self-governance. It is an area of innovation.

As a feature of cryptocurrencies this is far more important than it may seem at first. All other currencies implement a system, but the rules of governance are not usually stated explicitly and even if there are rules against debasing/inflating the currency—for example, the gold standard—then the governing body is happy to change the rules on the flip of a coin. With a cryptocurrency, the rules of governance are built into the software and enforced by the software. Consequently, the only way to violate the rules is to change the software.

There are already a variety of governance schemes implemented by different cryptocurrencies. Most schemes put the power to approve software changes in the hands of "stake holders" or "miners." Simply explained, a miner is someone who provides computer resources to run the blockchain and is rewarded according to some scheme when new blocks are created. Stake holders are similar, the difference being that they do not specifically provide computer resources, but they do deposit cryptocurrency that finances the provision of computer resources. They too are rewarded when new blocks are created, according to a specified reward scheme. The rewards that stake holders or miners receive are determined by the blockchain software and thus they are unlikely to approve any code changes that reduce their rewards.

An important governance issue is that if there are only a few stake holders (or miners), a group of them could form a cartel that represented a majority of mining resources. They could then take control of the blockchain code and, by changing it, defraud the users of the currency and other stakeholders. If that happened it would collapse the value of the currency and probably destroy it, but the cartel might be able to make sufficient profit from their fraud to justify the coup. Consequently, the governance mechanism needs to minimize or eliminate the possibility of such attempts to derail the governance of the currency.

When reading about blockchain governance, one frequently encounters references to "decentralization." Decentralization is the general term for governance mechanisms that prevent a majority of stake holders (or miners) from conspiring to defraud currency holders. In practical terms, if you have 1000 stake holders spread across the globe, it will be difficult for a majority of them to conspire effectively. If you have 2000 it will be more difficult still. If you have a mechanism that encourages currency holders into becoming stake holders, so that eventually millions of currency holders do so, then it will become almost impossible to corrupt the governance of the blockchain.

The Fiat v Cryptocurrency Comparison

A frequently discussed question is whether a cryptocurrency will eventually usurp all of the major fiat currencies, such as the US Dollar or the Euro. This is a complex question. To provide a coherent answer we need first to describe and then move beyond the experience so far with cryptocurrencies.

Ever since Bitcoin was a boy, most cryptocurrencies have experienced volatile price movements. Such volatility makes it impossible for these currencies to provide a reliable metric of value. Currently only a few cryptocurrencies are supported by a business model that is mature and has seen significant adoption. So at the time of writing few cryptocurrencies achieve their value from the strength of the underlying business model. As more of these cryptocurrencies mature there will be more cryptocurrencies with less volatile prices.

These cryptocurrencies may be better thought of as securities, like stocks

Factor	Fiat v Crypto
Medium of Exchange	Local fiat currencies are better mediums of exchange because more people readily accept them and they are easier to use. This may change.
Metric of value	Local fiat currencies are the de facto metric of value. A cryptocurrency that links to a fiat currency would have the same status. Most cryptos are too volatile to be a good metric of value.
Portability	Fiat currencies are slightly more portable that cryptos, but crypto can be almost as portable via cards and phone capabilities
Operating cost	Crypto has far lower transaction costs than fiat.
Fraud	Fiat is subject to greater fraud than crypto.
Divisibility	Crypto offers far greater divisibility than fiat.
Governance	The governance of fiat demonstrates regularly that it cannot be trusted. Crypto governance is far superior.
Coverage	Local fiat can be **bearer money** and also **fractional reserve money** as well as **account money**. Crypto can only be **account money**. Fiat is not international, but crypto is.

Table 3. Traits of currencies compared

or bonds. They are investment vehicles, designed for appreciation. As such they can never be true currencies. They can be a store of value, but like stocks and bonds, they will not become be a metric of value.

All the main cryptocurrencies that are intended to be currencies: Bitcoin, Bitcoin cash, Litecoin, Monero, Dash and Zcash, have been far too volatile to replace any fiat currency. We can blame immaturity (you just wait they'll be great currencies one day) or poor design (don't hold your breath they were doomed from the get go). In truth, it is possible to design a stable-coin, simply by linking it to a fiat currency.

So having noted all of that, let us now compare cryptocurrencies to fiat money, as summarized in *Table 3* above. We notice several things, which we can summarize in a single sentence:

Stable cryptocurrencies are a far superior currency of account for reasons of:

- much lower cost
- much lower possibility of fraud
- far better divisibility

– almost bullet-proof governance.

Because of that, at some point in time, when cryptocurrencies become much easier to use, they are very likely to replace fiat currencies as the currency of account—possibly just a single cryptocurrency will do this globally, but maybe it will come down to different cryptocurrencies in different geographical areas. Much depends on how governments respond to cryptocurrencies and how they try to regulate them or compete with them.

The Proportions of the Three Types of Money

A cryptocurrency could replace fiat currency, but only in respect of the *account money* in the economy. Bearing that in mind, it is interesting to identify the proportion of each of the three kinds of money within an economy. If we consider the situation just for the US dollar, using the figures published by the Federal Reserve Bank of St Louis, the amounts are as follows:

1. **Bearer money:** $1.69 trillion coins and dollar bills in circulation.

2. **Account money:** $2.01 trillion (calculated by subtracting M0 from M1 figures for October 2018)

3. **Fractional reserve money:** $14.1 trillion (M2 figure for October 2018—note that the M3 figure for October 2018 is slightly less than the M2 figure)

These figures show *account money* and *bearer money* combined as 26% of the US money supply. *Account money*, the proportion of the US money supply that cryptocurrencies could replace is roughly 14% of that total. Consequently, we conclude that there is no possibility that crypto-currencies in their current form could replace the dollar or any other national currency.

Nevertheless, cryptocurrencies could become a major force, especially if one becomes dominant globally. The total world money supply, expressed in US dollars for October 2017 (figures taken from The Money Project), is estimated to be $90.4 trillion. If we assume that 14% of that total were replaced by a dominant cryptocurrency, it would represent $12.6 trillion in value.

In the final chapter of this book we will pursue this topic further and examine the likely impact of cryptocurrencies on the banking world.

Chapter Summary

This chapter provides a full definition of what a currency is, then compares cryptocurrencies to fiat currencies:

- The seven characteristics of a currency are as follows:
    - > A currency is a medium of exchange.
    - > A currency must be portable.
    - > A currency must be inexpensive to create, maintain and use.
    - > A currency must not be easy to forge or steal.
    - > A currency must be easily divisible.
    - > A currency must act as a metric of value.
    - > A currency requires governance.
- The advantages that fiat currency has over cryptocurrency over are as follows:
    - > Local fiat currencies are better mediums of exchange, because more people readily accept them and they are easier to use. This may change.
    - > Local fiat currencies are the de facto metric of value. A cryptocurrency that links to a fiat currency would have the same status. Most cryptos are too volatile to be a good metric of value.
    - > Fiat currencies are slightly more portable than cryptos, but crypto can be almost as portable via card and phone capabilities.
    - > Local fiat can be *bearer money* and also *fractional reserve money* as well as *account money*, crypto can only be *account money*.
- The advantages that cryptocurrency has over fiat are as follows:
    - > Crypto has far lower transaction costs than fiat.
    - > Crypto offers far greater divisibility than fiat.
    - > Fiat is subject to greater fraud than crypto.
    - > Crypto is international, whereas fiat is not.
    - > The governance of fiat demonstrates regularly that it cannot be trusted. Crypto governance is far superior.
- *Account money* makes up roughly 14% of the US money supply. Insofar as any cryptocurrency could replace the dollar, it could only replace 14% of the supply of dollars.

The "Common Sense" of Cryptocurrency

Chapter 7

A Technology of Transparency and Trust

No problem can withstand the assault of sustained thinking.

~ Voltaire

⸺∞⸺

The most valuable transaction in a non-barter economy is the payment transaction. Someone pays money to someone else for something. Money moves in one direction and the something (some good or service) passes in the other direction. The buyer and seller both get what they want. But for this to happen, the money has to change hands. What we refer to as "the payment transaction" is the money changing hands.

Consider the simplest situation. Two people meet, one has **bearer money** and the other has goods. They swap and it's done. At least it seems that way, because each walks away with what they expected. But even that situation is not exactly simple.

The one who brought money had to get it from somewhere and the one who received money must now put that money somewhere. As **bearer money** was involved, some authority (nowadays, the central bank) had to issue the money and guarantee the value of it. It had to pay the manufacturing cost of the money and manage the system that circulates money.

If payment technology (debit card, credit card, mobile phone payment, or whatever) had been involved, then the receiver of the money had the technology required to receive the payment and they will (in the US) most likely pay a direct transaction charge for the payment. It would be the same if it were a cryptocurrency payment carried out directly between two wallets. There would be a transaction cost, although the transaction cost would most likely be lower and the banking system would not be involved.

The Transparency of the Blockchain

Payments made with **bearer money** are anonymous. Neither the government nor any other organization or person can know for sure that any transaction happened. Understandably, governments dislike this because they levy taxes on transactions (either sales tax or income tax). They get to know about **bearer money** transactions only when the money involved is taken from an account or put into an account that they can track.

So black market transactions go untaxed. Government money laundering regulations try to discover black market activity by analyzing reports that banks make of unusual amounts of **bearer money** passing through accounts. Businesses like casinos are natural vehicles for money laundering, because anomalous money laundering payments can pretend to be roulette wheel winnings.

When Bitcoin first started to hit the headlines, it was unfairly criticized for being "criminal-friendly." It is not. The Bitcoin blockchain is a readable public ledger. Movement of cash between wallets can be tracked, and if any money tries to leave the cryptocurrency system by way of an exchange into fiat (e.g. dollars or euros), it can be traced. Bitcoin transactions are transparent, but the names of the owners of the wallets are not public, except on exchanges where cryptocurrency is traded back into fiat. The transparency of transactions is fundamental to some cryptocurrency applications.

Bitcoin transactions do not have to be financial. Because 80 characters of space is provided for comments or reference information you can use the Bitcoin blockchain to store information. It is not an efficient way to store data, but it is an immutable incorruptible store. You can also encrypt the text field to obscure its contents, so you could, for example, store a very large data file somewhere, privately, and simply store its encrypted address on the blockchain. Nobody would know what it was except you. A Bitcoin transaction can thus be partially transparent (showing the amount) and partially encrypted.

The Versatility of the Blockchain

Some developers thought it a good idea to create a cryptocurrency where the transactions were completely private; untraceable rather than traceable. Soon several were created: Monero, Dash, Zcash and others. The upshot is that if you want to cover your financial cryptocurrency traces, you can, using these coins. For example, if you have 5 Bitcoin and then exchange them for Monero and then use that Monero to buy something, it will be impossible for someone to prove what you did.

The important point to note here is not that privacy can be guaranteed, it's that the blockchain has been used to create a completely new kind of currency: **account money** whose privacy can be guaranteed.

There is already a wide variety of cryptocurrencies. It is not difficult to invent and build new ones, since a cryptocurrency's behavior is programmed in. You can choose any characteristic you want: a constant invariable supply, a money supply inflation rate of 2 percent per year, or 10 percent per year, or one, like Bitcoin, where the rate of inflation gradually decreases. If you want one whose value is tied to a basket of fiat currencies, you can create one. If you want one that can only buy movie tickets, you can invent one. In general you can add any feature you want to a cryptocurrency and program it in. At the time of writing, over 2000 cryptocurrencies have been implemented, and many more will be created. Only a few, however, have become successful.

Governance Innovation

The vast majority of cryptocurrency software is released as "open source"—meaning that the program code is freely shared according to an open source contract. This has shrunk the software development cost for creating new cryptocurrencies, as you can take the base code of another cryptocurrency and change just a portion of it to invent something new. This has not only spawned new currencies, it has also spawned the new schemes for governing a blockchain.

Blockchain technology is still quite young and there challenges still to be resolved. For example, Bitcoin is governed by what is called "Proof of Work." The process is as follows:

> *The work to add a new block to the blockchain is distributed across a large number of competing "mining" computers, each of which tries to add the next block as fast as it can by doing work—trying to solve a hard cryptographic problem.*

As Bitcoin evolved and the price rose, the financial incentive to be the successful miner increased and computer companies built ever more efficient hardware to solve the cryptographic problem. By 2017 the cost of the electricity required to run Bitcoin mining was becoming extreme. At the time of writing the Bitcoin mining process consumes 2.6 GWatts of power—equivalent almost to the electric power needed by a country the size of Ireland.

Despite the fact that it has been possible to expend that much energy and money on mining, because of the level of reward, it is a horribly inefficient and expensive way of creating new blocks. Consequently, other solutions have been dreamed up, developed and implemented. The governance problem specific to Bitcoin is that Bitcoin mining involves many computers running a great deal of code to create a single block. Alternative solutions involve the expenditure of far less electricity and go by other names, such as "Proof of Stake" or "Proof of Burn" or "Proof of Authority." Blockchain innovation is proceeding and some blockchain issues that are reported in the press as problems are being solved.

Smart Contracts

When you read dramatic claims about the potential of blockchain technology the term "smart contract" is usually a big part of the picture. Up to now we have discussed blockchain technology as a practical system for establishing a currency that can act as **account money**. We have discussed variants of currency systems. Smart contracts add a great deal of capability beyond that level. If you think of the blockchain as a database of a kind, then you should think of smart contracts as the applications for that database.

The idea of smart contracts is not a recent one. The idea was invented and the term "smart contract" coined by Nick Szabo, a computer scientist, legal scholar and cryptographer. Szabo was investigating how to bring the sophisticated practices of contract law into electronic commerce protocols between parties on the Internet. When the blockchain emerged, it soon became apparent that smart contracts were an obvious complement.

A smart contract is a software module that can be attached to a blockchain which autonomously executes a specific function. Readable and executable program code is stored on the blockchain making it possible for anyone to examine it and see what it does. As the blockchain is immutable, once a smart contract is added to the blockchain its code can never be changed.

Consequently, smart contracts need to be tested exhaustively before they are implemented.

A simple example of a smart contract is one to manage the mail-order sale of goods between two people. The buyer does not want his payment to be made until he has received the goods and checked them. A smart contract can be created where the payment is made in escrow, and the money is only released when two out of three parties agree that the payment should be made. The three parties are the buyer, the seller, and an arbitrator. If the buyer is happy, the money is paid; if not, he will return the goods and reclaim the money. However, the seller receives the returned goods first. If the various steps involved (proof of delivery and return) are automated, the smart contract can be the arbitrator, deciding when to release the money and to whom.

Escrow of money (or documents) is an obvious and useful application of smart contracts. Because smart contracts are limited only by what software can do, you can create smart contracts for much more involved and ambitious applications. You could conceivably have smart contracts for complex insurances, large shipping contracts, or large building projects.

No matter how complex they may seem, contracts are agreements with clauses, dependencies, riders, and penalties. Usually all such details can be expressed in program code. The challenge is in finding ways to make such contracts self-executing and self-enforcing. Where you can do that, there can be a considerable cost saving over implementing the contract in a conventional way.

At the heart of smart contracts is the idea of automating trust. This is a powerful concept. In many areas of the modern economy organizations exist for the specific purpose of providing trust. And in many instances one encounters examples of the betrayal of trust, whether it's insider information, rigged elections, financial scams, or biased umpires and referees. The weak link in the chain is the human component. With smart contracts that weak link is removed. The smart contracts behave as programmed. They are immutable and trusted.

The Intermediaries

Smart contracts will make most impact in intermediary businesses like those listed below, either by making them more efficient or by spawning lean and efficient competition:

- Insurance companies are intermediaries that pool risk across a population of people or companies that wish to hedge against specific events.

- Financial markets are intermediaries that connect buyers and sellers of a wide variety of stocks, bonds, derivatives, commodities and so on.

- Shipping companies are supply chain intermediaries.

- Advertising channels intermediate marketers and customers.

- Retailers, whether bricks and mortar or Internet operations, bring together buyers and sellers.

- Recruitment companies are intermediaries that bring together those who seek jobs with those who are able to offer them.

- Search engines are intermediaries that bring together those seeking information with those who can provide it.

- Social networks are intermediaries that bring people together to share social information.

- Ultimately politicians are intermediaries between the voters and the machinery of government organizations.

Cryptocurrency projects have been launched in all these areas, and thus some incumbents may eventually face unexpected challengers. Others may simply adopt blockchain technology and adjust their businesses accordingly. Smart contracts are part of the picture, but not the whole picture. Because of the blockchain the world of data is also undergoing evolution.

The Blockchain as a Shared Database

For decades we have built computer systems that centralized data in databases that corresponded to the needs of departments or sections of the business. Computer systems were rarely built to share data between organizations, and as a consequence there are very few databases of that kind. The main reason for that was the problem of sharing such data securely—a problem the blockchain solves.

A blockchain is a secure shared database, not managed by any specific organization, but by the decentralized servers that provide the resources to operate it. Multiple organizations can use a blockchain to securely share transactional data. The transactions on the blockchain cannot be compromised because they are immutable. The data can be shared publicly (i.e. anyone can look at it), or the data can be encrypted so that only the holders of the encryption key can view it.

This set up is particularly useful for supply chain applications. Many of the reconciliations that are necessary in supply chain applications simply melt away when a shared transactional database is implemented. The longer and more complex the supply chain, the more useful blockchain technology will prove to be. Smart contracts can make it even more productive. They can be used to automate the interactions between the supplier and receiver, the transport companies involved, the insurers, customs, financing and so on. Ideally all of the parties would be able to view the data and to know the exact status of any supply chain operation.

Credit Scoring and Dirty Data

When Equifax was hacked in early 2017, it shone a light on the issue of personal data. Equifax, one of the world's three major credit reporting companies, collects and stores personal information from banks, mortgage servicers, debt collectors, and other credit providers. It then aggregates and analyzes the data, so it can sell credit reports and a FICO score to lenders.

The news headlines declared that the data of approximately 143 million U.S. Consumers, in addition to that of some Canadian and British citizens, had been compromised. It was particularly serious because the stolen data included names, Social Security numbers, birth dates, addresses, driver's license numbers and credit card details—data that enables identity theft.

The EU's data protection legislation (GDPR) was not in force at the time, so the company escaped the regulatory consequences it might otherwise have faced. Nevertheless the data heist came as a shock to those who didn't even know that Equifax had copies of their personal data. Some of the reporting that attended this high-profile hack drew attention to a disturbing aspect of the credit reporting industry—the data errors it makes.

According to the *Report to Congress Under Section 319 of the Fair and Accurate Credit Transactions Act of 2003*, over one in five consumers has a "potentially material error" in their credit file which accords them a lower credit rating than they deserve. Yes, that's over 20 percent. The consequences are that they suffer less favorable credit terms, higher interest rates or outright denial of credit.

Until recently it was notoriously difficult to get your credit rating changed if it was inaccurate. Under GDPR, the need for organizations to cater for the ability to change data has likely forced the credit reporting companies to improve their systems.

Credit rating companies are already facing competition from blockchain based start-ups. One, called POINTS, was founded in 2017 in China. It is building a credit scoring protocol on top of the recently-launched Ontology blockchain and it has raised $8 million in seed funding from a mix of traditional Chinese venture capitalists. It is backed by Zhong Cheng Xin Credit Technology, China's first nationwide credit rating agency. The leading credit checking companies will face blockchain competition sooner or later.

Dirty Data

Dirty data is a perennial problem with computer systems. You can probably guess the leading causes: incorrectly entered data values, data values not kept up-to-date, validation failures, software design errors, program errors, and hardware failure, including corrupted media. The reported level of error in credit scoring should cause us to pause and ask, "How does that happen?"

The findings of the congressional report only focused on the 20 percent plus whose credit scores were incorrect and detrimental. No doubt there will have been some examples where the credit score was better than it should have been. So it seems as though a good deal of data the credit scoring companies collect must be dirty. But that need not be the case. If only a small proportion, say one percent of the data, was incorrect, it could affect the accuracy of many credit scores.

The problem resides in the fact that the collected data has many sources and that the subject of the credit score—who has a great interest in it being accurate—is not involved in the data gathering and validation process. In short, the conditions for establishing trust in the process are absent, despite the fact that all the data being processed is personal data. If all the data were held on a blockchain and the process of calculating a credit score was an auditable smart contract, it would be different.

It is our expectation that as blockchain technology proliferates it will, aided by data regulations, improve the dirty data situation, simply because it will expose more data to the light of day.

Data Ownership and Data Sharing

In *Chapter 3, The Data Rights of Man*, we discussed personal data and proposed rights individuals should have over it. We need now to consider all other data. The data not owned by the individual is owned by

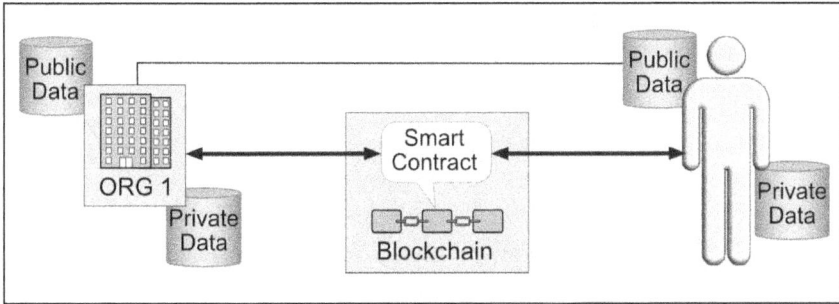

Figure 6. A Smart Contract for Data Access

government, businesses, or other legal entities. All such organizations have identities—official credentials that relate to their creation and their existence. Although such data can be voluminous, it is very similar to personal data, both in structure and the legal status it should be accorded.

With these ideas in mind, we illustrate a simple smart contract-driven data interaction between an individual and an organization in *Figure 6*.

We show both an organization and an individual, in each case classifying their data as either public (the data owner is willing to share such data) or private (the data is not intended for sharing at all). The organization and the individual agree, by means of a smart contract, to provide the organization with access to a specific subset of the individual's public data.

The contract will include constraints. For example, the organization might be given permission to read the data but not store any copies of it. Alternatively it might be given permission to store the data but only for a given time period or to store the data but only use it in specific ways. The contract could insist on an audit trail of the use of the data and specific penalties if the contract is violated.

Now consider various common situations where data usage could be organized in this way.

- Social network data—comments, chat, photos and so on—could be constrained so that it could only be accessed by specific people under specific circumstances.

- Personal credential data—driving license, passport, social security number and so on—could be made available as proof of identity, but might insist on being retained only for a day.

97

- An email provider might be constrained to never read the emails they deliver.

- A voting machine could be constrained to provide an audit trail of your vote in an election to prove that it was recorded and that it was counted.

- Your complete health records could be made public on request by any health care professional with the appropriate credentials, making the data immediately available in any medical emergency, but providing discretional access at all other times.

- Content creators (authors, directors, musicians, artists) could sell their digital products subject to data constraints, such as never allowing them to be copied, or even only allowing them to be used a specific number of times.

The Zero-Knowledge Concept

Although complex and perhaps a little difficult to understand, zero-knowledge proofs provide a very useful technique to the cryptocurrency world. By definition:

A zero-knowledge technique is one where one person can prove to another that they have specific knowledge without revealing the knowledge.

Consider the situation where you are entering a secure area and the guard on the gate wants to know that you have security clearance before giving you access. With a zero-knowledge proof, you could prove to the guard that you indeed have the necessary credentials without even showing a security card. Remarkably, this is possible.

Expressing it more mathematically:

With a zero-knowledge proof the "prover" can prove to a "verifier" that he/she/it knows the value of a particular variable, say x, without conveying anything to the "verifier" other than that the prover has this knowledge.

So let us say that the value of x is 42. The challenge for the prover to prove he/she/it knows that value without passing on any information at all that relates to the number "42."

A zero-knowledge proof satisfies the following three properties:

1. **Completeness:** The proof must convince the verifier that the prover has the knowledge.

2. **Soundness:** The prover can only convince the verifier by passing correct information. (In other words, if a prover passes any false information at all it invalidates the proof).

3. **Zero-knowledgeness:** No information of any kind relating to the knowledge is passed during the proof.

An Example: The Color Blind Man

Consider the following situation. A color blind man has in his possession two colored balls, one red the other green. As he is color blind he cannot distinguish between them and does not even know for sure that they are differently colored, or different in any other way. To him they appear identical. You wish to prove to this man that you know that the balls are different. However you are not allowed to tell him what color each ball is. In fact you are not allowed to tell him anything about these balls.

The color blind man is the **verifier** and you are the **prover**, who must prove you know something without revealing any of the detail of what you know. Here is how it is done:

You pass the two balls, ball "A" and ball "B," to the verifier who keeps them out of sight in boxes labelled "A" and "B." He takes one of the two balls, from the box labelled "A" and shows you it. He then returns the ball into the box. He then randomly chooses one of the two balls, takes it from the box and shows it to you. He knows whether it is ball "A" or "B," from the label on the box he took it from. You know which it is by the color. So when he asks you whether it is ball "A" or "B" you tell him. The procedure can now be repeated as many times as the verifier wishes and in each instance, you, the prover, will be able to identify which ball is which, but the verifier has now idea how you do it. With enough repetitions the verifier will conclude that you are able to distinguish between the two balls.

If the number of verifications carried out is 10, then the odds of guessing the balls correctly by chance would be about 1 in a thousand. With 20 verifications it would be one in a million.

The zero-knowledge proof idea is not new. It was invented in 1985 by Shafi Goldwasser, Charles Rackoff, and Silvio Micali. They created the notion of "knowledge complexity," a metric for the amount of knowledge needed for a prover to pass to a verifier to convince the verifier that the prover possesses a specific piece of secret information. They were able to prove mathematically that with some interaction between a prover and verifier the amount of secret information that needed to be transferred was zero. So, zero-knowledge proofs are based on sophisticated but well documented mathematics.

As time passed mathematicians added more sophistication. It was proved that any interactive proof system (i.e. one with a verifier and a prover) could be proved with zero-knowledge. Building zero-knowledge proofs that could work over the Internet was a further challenge that was resolved with witness-indistinguishable proof protocols. Finally, non-interactive zero-knowledge proofs were invented, so that there could be proofs where no interaction between the prover and verifier was required— only a common reference string shared between the prover and verifier was needed to achieve computational zero-knowledge.

If you think this sounds like magic, you are not alone; zero-knowledge proofs are sometimes referred to as "crypto magic."

Applications of Zero-Knowledge.

The importance of zero-knowledge proofs cannot be overstressed. When we discussed different kinds of data in *Chapter 3* we noted that there were only three different kinds of personal data:

1. Credential Data

2. Personal State and History Data

3. Title Data

For all practical purposes, the second and third type of data can be controlled by the owner and revealed only with the owner's permission. Credential data, however, is required to be revealed in many circumstances and thus it can be difficult to prevent such data being copied and exploited by others. Zero-knowledge proofs make the issue melt away like snow on water. You can use them to prove that you have credentials without needing to reveal the credential data. They can be used to prove that you know a password without need to reveal the password. In general, they can be used to verify identity without revealing any identity information. This is their most important application.

Another application of zero-knowledge proofs is their use by several cryptocurrencies to anonymize payment transactions. They can be used to obscure all the details of a transaction: sender, recipient, amount and anything else the transaction contains.

One example of the application of a zero-knowledge technique is Monero's use of a ring signature. A ring signature is a digital signature that can be executed by any member of a group of users that each have keys. When a transaction is signed with a ring signature, you can prove only that one of the group signed it—it is impossible to know which member of the group it was. Monero uses ring signatures to hide the identity of the initiator of a payment transaction.

The anonymization of payment transactions has far reaching implications. Before the creation of these cryptocurrencies, the only way to achieve anonymous transactions was by using coins and notes. Governments are unlikely to be happy with this innovation if it sees extensive use, because it assists tax evasion. However, it is difficult to think of a way to stop it. You can legislate against its use, but if you cannot identify who is using the currency, the legislation will be toothless.

The Interplanetary File System

Given that most personal data is widely scattered across many storage locations, on cell phones, tablets and personal computers, on websites like Google, Facebook and many others, it is natural to wonder how it can all be consolidated in a way where it is possible for individuals (or organizations) to manage it.

A good deal of work has been done on this in academic circles and more recently within the blockchain community. For the benefit of the non-technical reader we will describe it without diving in to the technical details. We can start with the fact that a file system is a means of storing data files and making them quickly available for access. PC and Mac users understand a file system as the way data is stored on their computer. That's a specific kind of file system. It consists of a directory with folders which can contain both files and other folders, which can also contain files or folders, in a hierarchical structure.

There is another file system which we all use, which is far more sophisticated and which changed the world: the HTTP protocol.

The difference between this and PC file systems is that the HTTP protocol presides over a distributed file system. The files to which it provides access are scattered over millions of computers all over the world.

You may think that the HTTP access command you put in your browser (e.g. *http://www.strangewords.com/news/blogs/090718*) is calling up a web page. In reality it is just pointing to the location of a file. The file itself contains both the instructions on how to format the web page and the data that will appear on the web page. The first part of the command (*www.strangewords.com/* in this example) identifies the specific computer where the file resides; the second part (*news/blogs/090718*) names the file. What enables this file system function is a DNS system, which looks up the physical address of the web server.

The HTTP protocol has served the world well, but it has defects. IPFS has been specifically designed to remove many of those defects so that ultimately it could replace it. To give you a sense of it, we'll list its capabilities:

- It is a peer-to-peer arrangement rather than hierarchical. This makes it far more efficient in its use of network resources.

- It is faster at and more efficient in transferring very large files. It uses a mechanism similar to BitTorrent for transferring large files.

- It places data close to where access requests for it are likely to be made. This and the previous two points means that it is far lower cost and also more capable than HTTP.

- It completely secures the data it stores (it is immutable) and also naturally deduplicates data. (It will keep multiple copies for redundancy but it eliminates excessive copies.)

- It has a content-addressing capability that allows files to be accessed by their content.

- Every file can be found by human-readable names via the decentralized IPNS (InterPlanetary Naming System).

The important role IPFS could fill is that of providing you with a personal directory to oversee and manage all your data. Technically it doesn't matter where the data is physically located, it only matters that it is secure, immutable and can be accessed from a directory that you control. IPFS provides that service and decides where to place the data itself.

From our perspective, a file system like this is also desirable because it can be used to store data that knows what it is and it integrates very easily with the blockchain. Files are identified by a hash key, which can easily be stored as part of a transaction or as part of a smart contract, on a blockchain. It is too early to say whether IPFS will become the dominant file system of the future. Nevertheless, in the process of technology evolution, it, or something like it, will inevitably replace the HTTP protocol in time.

The Solid Project

Prof. Tim Berners-Lee, the inventor of the web in its current form, is unhappy with how it has developed. He began work on the project called Solid back in 2003. The goal was to fix many of the Internet problems that had begun to emerge by then. (SOLID is an acronym of SOcial LInked Data.)

Berners-Lee's original vision was that the web would become a massive, decentralized, collaborative "read-write" space. The first browser, which was called WorldWideWeb, had a read-write capability, able not just to read web pages but also to edit them. However, despite the original intentions, the web centralized. Business Behemoths like Yahoo, Google, Amazon, Facebook and others, seized the lion's share of commercial ownership, riding on the back of accumulations of personal data.

Project Solid is an attempt to reverse this movement towards centralization—to decentralize the web, but do so in a way that is compatible with the web as it exists today. It consists of a technology stack—a group of related protocols.

To understand how it works, consider the fact that nowadays we have multiple devices: mobile phones, tablets PCs, possibly at home as well as at work. The possibility of your data being stored in a single place now seems impossible, as some of it is likely managed by a cloud storage application like Dropbox and some resides in the large data centers of Google, Amazon, Facebook, et al. The coherence of your data thus depends upon various applications, some of which might run on your devices (such as email) and some of which are website applications like Google and eBay.

Every application manages its data for itself and provides no means of sharing any of it. There is no common data resource for personal data and no standardization of data structures. It can even be difficult to get access to your own data—although that problem has been reduced by the EU's GDPR regulations. The practical problem is that there is no personal data resource available to anyone and no easy way to create one.

Solid fixes that problem, making it possible for multiple applications to share a common resource for data storage and retrieval. It calls this data resource a "pod." In practice a pod is a set of links to data, with the actual data being distributed across multiple devices (your own devices, cloud storage and so on). In Solid, you store the data you produce wherever you please and access it using URLs, similar to the web. (This, incidentally, could be integrated with IPFS.)

Solid makes use of RDF, a semantically based standard for data interchange which is capable of representing any kind of data structure, including statements in English. This is one of the standards that make up the Semantic Web, the goal of which is to make it possible to connect and analyze all data.

A Blockchain-Centric Universe

Computer systems involve interactions between four distinct categories of things: people, software, data and computer resources. This is true no matter whether it's a simple program doing something trivial on your cell phone, or it's the whole of the world-wide web, buzzing away in all its real-time glory.

Blockchain systems can be viewed as the interaction of five categories of things, as illustrated in Figure 7—the previous four we identified, plus the blockchain itself. You could argue that a blockchain involves data, software and computer resources, and could be lumped in with those things, but the truth is that the blockchain does something that systems previously didn't do: it treats the transactions of a system as separate. It does so in a way that makes them immutable (and hence incorruptible) and (potentially) public (i.e. auditable and transparent).

If we consider the data storage ideas of IPFS and Solid, we can imagine a system where all the data is either transactional (and on the blockchain) or distributed (versioned and immutable) across many data stores that are linked to the blockchain. The software that runs the system can consist of smart contracts that are held on the blockchain and all the resources that run the system (computers and networking hardware) are distributed across the world.

Putting it all together you get a fully distributed system. The system is collaborative, in the sense that the resources can be owned by anyone and located anywhere, the blockchain itself doesn't need to be owned by any specific business or organization, the software can also be a collaborative

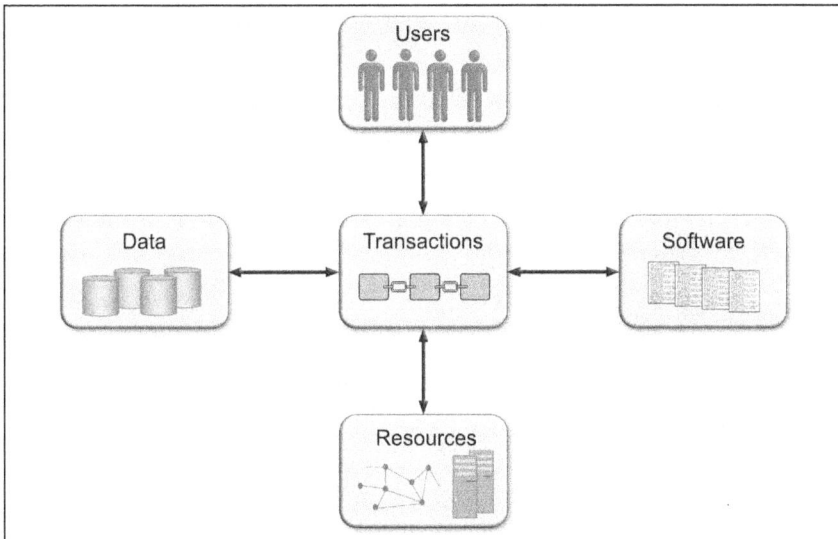

Figure 7. Blockchain-centric Computing

operation and all of the activity of the system can be financed by a cryptocurrency. Most importantly, the system can be built so that trust in the fairness of the system is embedded and ensured by its structure.

Existing businesses can take advantage of all these technologies and, where appropriate, use them to replace their existing systems or create new ones. It has begun to happen. IBM, for example, who is quick to jump on new technology developments, has plunged in with obvious enthusiasm, helping a variety of banks develop blockchain payment systems. Indeed, banking is the most active area, accelerated perhaps by the success of the Ripple payments system. Indeed at the time of writing, Ripple is the only cryptocurrency that has demonstrated a successful application. Adopted by 11 of the world's top banks and well over 100 in total, it provides a very inexpensive and lightning fast cross-border payments service.

Other examples of mainstream companies getting blockchain fever include:

- Samsung is building blockchain solutions for public safety and transport applications along with the South Korean government.

- Barclays (UK bank) building systems for tracking financial transactions, for the sake of compliance and combating fraud.

- Maersk has plans to streamline marine insurance using the blockchain.

- Kodak is developing a blockchain system for tracking intellectual property rights and payments to photographers.

- Reliance Industries (India's largest conglomerate) is developing a blockchain-based supply chain logistics platform along with its own cryptocurrency, Jiocoin.

- The Estonian government has partnered with Ericsson to move public records onto the blockchain.

All are all examples of businesses adopting blockchain technology, and not of businesses being constructed in a blockchain-centric manner. There is a significant difference between the two.

Chapter Summary

This chapter focuses on various aspects of blockchain technology and their capabilities.

- The blockchain, initially implemented by Bitcoin, is a readable public ledger that allows anyone to track the movements of all Bitcoin between wallets. The blockchain can thus be used to form an immutable and transparent record of transactions.

- The blockchain enables the creation of new digital currencies. This is facilitated by the fact that the bulk of cryptocurrencies are built on open source software. In our view, currently the most important area of innovation is in the governance of the blockchain and hence the underlying cryptocurrency.

- The blockchain can be thought of as a shared database. Smart contracts, which are stored on the blockchain, can be thought of as application software that runs on the blockchain. Smart contracts are inviolable.

- A very powerful idea that has become a component of blockchain technology is the zero-knowledge proof. A zero-knowledge proof is one where a person can prove to another that they have specific knowledge without revealing the knowledge. Such proofs have multiple applications, as they enable anonymization and can be used to check credentials without the credentials needing to be shown.

- There are two projects currently in progress which seek to rethink the way the internet works. They are not directly competitive and it is likely that both will gain traction:

    > The Interplanetary File System offers an alternative file system to the one defined by the HTTP protocol and has many advantages.

    > The other, project Solid, initiated by Prof. Tim Berners Lee, is an attempt to decentralize the web, but do so in a way that is compatible with the web as it exists today.

The "Common Sense" of Cryptocurrency

Chapter 8

Old Dogs, New Tricks

All mankind is divided into three classes: those that are immovable,
those that are movable, and those that move.

~ Benjamin Franklin

———∞∞∞———

Over a decade, from 1995 to about 2005, the Internet rearranged the world's economic landscape, forcing a wide array of businesses to adopt Internet technology and build "brochure" websites or even interactive websites. This new landscape was particularly difficult to adjust to for established brick-and-mortar retailers and traditional publishers of newspapers and magazines. Some fell by the wayside, others survived, and one or two thrived. But none who thrived became phenomenally successful.

The accolades and the glory went to the companies that re-imagined the business world in terms of the new technology that underpinned it: eBay, Amazon, Yahoo, Google, Salesforce.com, Facebook and others. It surprises no-one that some of these businesses are now among the most valuable in the world.

Mark Twain insisted that "history does not repeat itself, but it rhymes." And indeed, in the metaphorical sense, the blockchain rhymes with the Internet. These are the rhymes:

- **A business revolution based on a new technology.**

  In the case of the web it was the HTTP protocol, in the case of the blockchain, it is the blockchain itself, smart contracts and a changed global file system (most likely IPFS).

- **An epidemic of new start-ups.**

  The web spawned a large number of new start-ups and the blockchain has done the same. It is already clear that only a small percentage of blockchain start-ups will gain traction, just as only a relatively small number of websites did.

- **An investment bubble.**

The route to attracting investment has been distinctly different with blockchain businesses—ICOs rather than IPOs. Nevertheless, the speculative excess has been similar so far. Many dot-com companies vanished into bankruptcy around late 2000. If we assume that 2017 was the peak of speculative blockchain investment mania, it's clear that many of the early blockchain start-ups are already out of business or heading in that direction.

- **A market collapse**

In the dot-com decline, the NASDAQ lost about 80% of its value. At the time of writing, in the current blockchain decline, the cryptocurrency market has lost about 80% of its value.

- **Technology innovation.**

The dot-com revolution provoked a whole series of technology developments. It commenced by establishing the network and website-building software. It proceeded with web commerce and web search. After that it was cloud computing and big data. That trail of technology spanned about 20 years, and the road it carved out made sense best in the rear-view mirror. The same is likely to occur with blockchain technology. We know it is about more secure computing, immutable data, and decentralized operations, but we do not know where it will end up.

- **Different Business Structures.**

Businesses became less hierarchical with the advent of the PC, as it distributed personal computer power across the organization. The dot-com businesses were less hierarchical still, tending to outsource everything but the core activities of the business. Blockchain businesses are likely to push this trend to its extreme to create fully decentralized structures with innovative forms of management structure (i.e. governance).

Taken together, this suggests that the blockchain will bring considerable disruption, as its technology complements or supersedes the technology of the Internet era.

The question is: Where will the impact be greatest?

A New Payment Technology

Every financial transaction you make involves a cost of payment that is levied by the payment system used to complete the transaction. Most payment transactions use computer systems and networks that are quite old and have relatively high costs. For example, the fees for a $100 credit card transaction are likely to be between $2.50 and $3.00. Payment costs, whether for credit cards, debit cards or checks have hovered around this level for quite a while.

Blockchain payment costs are lower. A breakdown of transaction fees for some of the major cryptocurrencies in January 2019, gives the following:

Bitcoin: $0.284

Ethereum: $0.082

Litecoin: $0.019

Bitcoin Cash: $0.005

Dash: $0.013

For the sake of a fair comparison, we need to take several things into account. In cryptocurrency's favor is that most transactions clear in minutes, far faster than traditional payment transactions. However at the time these costs were recorded none of the currencies shown were being stressed by large transaction volumes. When they are, the transaction costs can rise by a factor of 10 or more, as some users bid up the cost for the sake of fast transaction execution. But even in such circumstances the transaction costs are usually much lower than for traditional payment systems. Blockchain technology delivers low cost payments and the costs will surely fall as as the technology evolves.

The Micropayment Dimension

As a consequence, new business ideas that need lower payment costs to thrive will naturally emerge as payment costs fall. This is particularly likely to have an impact on the consumption of digital media. Paywalls of the kind that have proved successful for big name media like The Wall Street Journal, The New York Times and The Washington Post, could be implemented by much less popular newspapers and magazines on an article by article basis. Bloggers would capitalize. Music streaming could be served up on a more granular basis, linking artists directly to their fan base. Interactive gaming is also a potential application area, where micropayments could support low cost in-game purchases.

A blockchain micropayment business called SatoshiPay entered this market in 2014 and has been gradually improving its capability. Its goal was to enable payments as low as a fraction of a cent. It used Bitcoin initially, but stalled when Bitcoin transaction costs suddenly escalated. In 2017 it began partnering with the Stellar Network to use the Lumens (XLM) cryptocurrency. The way that SatoshiPay works is that user funds are held in a user's browser and each individual transaction (for reading an article or whatever) is paid directly to the publisher. The payment mechanism works across websites and consumers can use the service without having to sign up or download software.

The ICO Phenomenon

The Initial Coin Offering (ICO) craze exploded in 2017. By the end of the year ICOs had reaped about $6 billion in investment, New ICOs were being launched at the rate of 50 new ICOs every month. In simple terms, an ICO is similar to an IPO (hence the name), selling quantities of a new token or coin, rather than stock, to the public. The tokens sold become functional units of currency if or when the ICO's funding goal is met and the project launches.

ICOs proved to be an excellent source of revenue for Ethereum in that year. Rather than just being a cryptocurrency, Ethereum is also a platform that other cryptocurrencies (called ERC20 tokens) can use as a blockchain. There are currently hundreds of ERC20 tokens that use the Ethereum blockchain in this way. Even though many of them are now defunct, those that survive constitute a useful application of the Ethereum blockchain.

Sadly, ICO activity in the US dried up after the SEC started questioning whether specific cryptocurrencies were securities or not. The critical point is whether a token is marketed as an investment. If it is, then it is a security and, as such, must abide by the regulations that cover marketing securities (for details, google the Howey Test). If it is not, then one might wonder why an ICO is taking place at all, since very few people buys coins at ICO without an expectation that they will increase in value. In any event, US-based ICOs are now as rare as hen's teeth. ICOs are hosted in more crypto friendly countries.

Cryptocurrency Investment: Publicly Transparent Businesses

The advent of cryptocurrency as a kind of investment is a wholly new financial development. You can think of it in this way: a cryptocurrency business does not need to have stockholders, and may not have them. There are no stock holders in Bitcoin or Ethereum or many other cryptocurrencies. However, you can buy a quantity of cryptocurrency as an investment in the hope it will appreciate. If so, you are assuming that the cryptocurrency itself will generate revenue.

Investors in stock can follow the fortunes of their investments through company news and quarterly financial reports. Cryptocurrency investors have no such information to peruse. However they can observe the trading of their beloved cryptocurrency in real-time by viewing blockchain activity itself and by using the various cryptocurrency analysis web sites: CoinMarketCap.com, BitInfoCharts.com and others.

In the future cryptocurrency businesses could, and may, set themselves up to be publicly transparent, so that all currency trades and the execution of all smart contracts on the blockchain are publicly available to examine.

Consider, as an example, the idea of a blockchain retail business that acts as an intermediary to bring together buyers and sellers of, say, collectible antique swords. It issues a cryptocurrency for trading purposes and uses smart contracts to govern all trades, collecting a small commission on every trade. If there is a fixed supply of the cryptocurrency then the value of the currency will increase according to the demand, which in turn will increase according to the frequency of transactions. The value of the cryptocurrency can be tracked and estimated by analyzing the blockchain behavior. If the active market for antique swords expands the currency will prosper. If it declines and dies the currency will become worthless.

A helpful analogy is to think about cryptocurrencies as if they were the economies of different countries. The value of a country's currency is ultimately determined by the foreign exchange markets, but it is reflected in the economic activity of the country and particularly by the balance of trade, since the trade in goods will be matched by sales and purchases of the country's currency. A net inflow of currency due to a positive balance of trade raises the currency's value and a net outflow lowers it.

Fully Public Companies

It is possible to envisage a fully transparent business or organization where every activity is governed by smart contracts and the payment for every transaction the organization makes is recorded on the blockchain. It would include every payment for all supplies, even office stationery, all services, all rent and insurance, every hiring or firing, all wages, and of course all income from sales. In effect the full accounts of a business could be reported in real-time. It is debatable as to whether full transparency would make good business sense—nevertheless, it is a possibility.

Businesses of this kind would offer a new approach to capitalism, where the cryptocurrency market substitutes for the stock market and the investor is informed about the business with real-time data rather than the stock market's traditional quarterly report. It might seem as though any business could be run in this way, but in practice it is probably not the case.

Because the governance of the blockchain and the supply of coins or tokens are determined from the outset and enshrined in software, blockchain businesses need to be fairly one-dimensional. For example, a blockchain business that sells sports and cinema tickets could work. More complex businesses like, say, shipbuilding or computer manufacture would be far harder to organize as a blockchain business, because of the diversity of their activities and the complexity of their funding needs.

Cryptocurrencies are commodity-like in this way.

Record Keeping Businesses

The blockchain has an unparalleled capability for keeping records, and there are many data items that require a secure immutable store that provides public or semi-public access. There are thus likely to be many new blockchain businesses or organizations that emerge in this area. Think in terms of people, property, organizations and regulation. For each category there are a multitude of documents that need to be stored:

- **People:** Birth certificates, passports, death certificates, voter IDs, contracts, signatures, wills, trusts, escrows, educational records, medical records, HR records, etc.

- **Property:** Land titles, property titles, vehicle registrations, product registrations, including digital products (video, music, games, software) etc.

- **Organizations:** Business incorporation/dissolution records, ownership records, board level records, licenses, permits, annual records, contracts, etc.

- **Regulation:** Permits, inspections, tax records, legal (court records, criminal records, regulatory records) etc.

Record keeping blockchain businesses will naturally arise in many areas and supersede the existing systems and organizations that store and manage such data.

Key Management Businesses

The current regime of password management for access to web sites and cloud capabilities and local devices can be improved significantly. The public key-private key arrangement can be made even more secure with two-factor authentication and applied to almost all situations where keys are necessary: the home, hotel rooms, vacation homes and time shares, cars, lockers, safety deposit boxes, all web sites, particularly those where money is involved such as bank accounts, investment accounts and betting.

The possibility of a single personal capability to manage all such keys is compelling. There are already software products that do much of this, but they could become blockchain based and be extended beyond their current capability.

Video Gaming

The size of the global video games market was about $137.9 billion for 2018, a full 8% of the media and entertainment market and growing like bamboo at 13% year-over-year. The blockchain is destined to disrupt this market as it can change the nature of what a video game is. It will allow players to earn cryptocurrency directly from playing games, or in the case of some multiplayer games, securely own and trade their virtual assets.

Most multiplayer games are pay-to-play, with gamers buying access and paying for game items: skins, weapons, powers, etc. Such games could be far more transactional, allowing gamers to trade items and potentially make profits—all driven by a cryptocurrency. If you attaching virtual assets to a cryptocurrency they become potentially more valuable, because they become more tradable. They might even migrate from one video game to another and thus exist in multiple game universes.

Naturally this expands the possibilities for the games themselves.

CryptoKitties was the first popular cryptocurrency-based game. It was simple and well thought out. Collectors (players) could buy, sell and breed CryptoKitties. CryptoKitties have a "genome" so that when two CryptoKitties are bred, the newborn inherits characteristics from the parents. To expand the gene pool, new CrytpoKitties with new characteristics were introduced into the "gene pool" every 15 minutes until November 2018. Since then new CryptoKitties only appear by breeding.

Every CryptoKitty is unique, owned by someone and stored on the blockchain. Its value is determined entirely by the market. CryptoKitties made the news in December 2017 when a CryptoKitty (the genesis CryptoKitty) sold for more than $100,000. They made the news again when the level of trading reached a point where it saturated the Ethereum blockchain, causing long delays in Ethereum transactions.

Other blockchain-based games include: EOS Knights, World of Ether, MegaCryptoPolis and Dragonereum. In 2018 alone about half a billion dollars has been generated in crypto games, so it is an undoubtedly healthy sector of the blockchain market, and it is still a nascent market in which virtual reality (VR) and augmented reality (AR) have yet to step into the ring.

Peer-to-peer Publish-subscribe Businesses

Blockchains operate in a peer-to-peer manner. Because of this they are, by nature, international—anyone almost anywhere can connect to and interact with anyone else, as long as they have access. The exception is countries where governments impose some form of control of the Internet.

Peer-to-peer networking is reflected, at the personal usage level, by a "publish-subscribe" approach to business interactions. A "publish-subscribe" interaction is where one person publishes information into a shared data space and other network users choose to subscribe or not to that information service. An example of this is a web site. More recently social networks provide examples of this kind of interaction.

Facebook (and some other social networks) violate the publish-subscribe dynamic by pushing unsolicited content (ads mainly) to its users, but the main activity is publish-subscribe. The ride hailing company Uber and the room rental business of Airbnb are both publish-subscribe businesses. The Google search page could be viewed as a publish-subscribe directory where users seek out web pages or web sites to subscribe to. Almost all open trading markets and online brokerages work in a publish-subscribe manner, as does eBay, the online auctioneer. All of these businesses will find themselves in competition with blockchain start-ups, if that hasn't already happened. Uber, for example, already has competition from Arcade City and Airbnb from Bee Token.

Whether the blockchain competition gains traction will depend on whether it can offer a superior proposition to the users of the existing Internet businesses. This in turn will depend on a variety of factors. To outcompete Uber, for example, you probably need to offer drivers a more rewarding proposition and yet offer riders a better deal. You also have to match the powerful analytics that Uber uses to run its operation. The advantage that a blockchain approach may be able to bring to bear in this situation might be to make drivers (in the case of Uber) stakeholders in the blockchain cryptocurrency, or renters in the case of Airbnb. Most likely, blockchain competition will arise against entrenched publish-subscribe businesses, by turning some of the participants into stake holders.

Countries that police their citizens (and other country's citizens) through covert electronic surveillance of their digital activity, may be unenthusiastic about blockchain technology—or at least their three-letter-agencies will be. However, they will probably find it difficult to oppose a business revolution, as politicians will fear losing out to foreign competition.

The Personal Freedom of the Blockchain

Right now people have very little control of their data, and consequently they have no means of using it as a resource. However, as this changes, an individual's ability to use their data as a resource will increase. It is worth summarizing the current situation:

An individual's basic credentials: their birth certificate, passport, driving license, social security, and so on, are conferred upon them by governmental organizations. Their educational record and employment history is similarly fragmented, with the data being held in multiple places. Their financial information is not under their control. Their health information is not under their control. Their record of ownership (of property, car, insurances, etc.) is not under their control. Their social information is not under their control.

The various organizations that control this information have their own interests, which do not necessarily coincide with the best interests of the individual. This is what needs to change most, and it will. It is critical for people to come to understand that their data has value.

From one perspective it obviously has value, because so many data

117

aggregators, from Google to the government, extract value from it and exploit it, in most cases without the consent or even the knowledge of the data owners.

So let us now discuss data value.

Consider money. **Bearer money** of the paper variety is in practice a "title to value." In other words, it is accepted that whoever carries it is entitled to the value printed on the note and has the right to exchange that value. Money held in bank accounts is also a "title to value" and exchangeable. So are investments. Any digital record that proves you own something, a computer, a TV, furniture or whatever, is also a "title to value." Other digital data is also "title to value," such as: written works, art, music, video and games.

All of these things are exchangeable, just as money is, with the proviso that a buyer needs to be found, the associated item needs to be delivered, and because the price of such things varies, a price needs to be agreed.

With the continual expansion of the world's computer resource and the software that runs on it, a massive amount of data has found its way into the digital world, and a good deal of that data has exchangeable value.

As data becomes standardized, as it inevitably must do, it will be possible to write smart contracts to enable easy and fast semi-automatic trading of this value. The possibility is arising to dramatically extend the markets that the likes of eBay and Amazon preside over.

The Cryptocurrency Banking Sector

Currently, there is a variety of bank-like activity that occurs in the cryptocurrency sector. Some cryptocurrency exchanges, Bitfinex and Poloniex are examples, make cryptocurrency loans to traders. This activity involves matching a borrower to a lender and charging a fairly high interest rate to the borrower, of which the exchange takes a cut. It is similar to a depositor putting money in a savings account for an agreed period and the bank loaning that money directly to another person. Such loan agreements could be defined by a smart contract, but in most instances this activity simply implements clearly stated rules of operation.

In this way cryptocurrency holders that have no intention of selling, can earn interest on their holding. It carries the risk that the borrower may default at some point. But the risk of this is low. Usually, the worst consequence of borrower default is that the lender forgoes the interest payment for a while; the cryptocurrency deposit itself is not at risk and the exchange will usually compensate the lender for lost interest.

There are some cryptocurrency businesses that make loans in fiat currency against deposits in cryptocurrency. If the value of the cryptocurrency collateral falls sufficiently, the loans they made will exceed the collateral they hold. If that happens they can fail, just as banks which make too many bad loans fail.

In our discussion of money, we divided money into three distinct categories: *bearer money*, *account money* and *fractional reserve money*, noting that cryptocurrency in its current form was only capable of being *account money*. There are no crypto businesses involved in the equivalent of fractional reserve banking, because cryptocurrency systems do not allow the creation of *fractional reserve money*.

Cryptocurrency and Fractional Reserve Banking

Fractional reserve banking depends on there being a "lender of last resort" to stabilize the banking system if one or more major banks get into trouble. We witnessed such a banking event when, in September 2008, Lehman Brothers filed for bankruptcy, declaring $639 billion in assets and $613 billion in debts, making it the largest bankruptcy filing in U.S. history. The value of Lehman's assets were collapsing as property values dived.

The impact rippled through the financial systems, causing an immediate liquidity crisis. In effect there were runs on many banks, putting them in danger of collapse. If there had been no "lender of last resort," many banks would have failed and others, in a stampede for liquidity, would have called in loans made to financially healthy borrowers, giving the US economy a heart attack. the banks needed liquidity and the Federal Reserve provided it. There was a dramatic increase in the US banks, call for excess reserves to meet the crisis, and the Fed added $813 billion to those reserves – an increase by a factor of about 19 – to meet the demand. The banking system was over-extended.

With the fall of Lehman a huge amount of *fractional reserve money* simply evaporated, collapsing the money supply in the economy. So the Fed printed money to make up the shortfall for the banks. This did not forestall a severe recession, as you may remember. The *fractional reserve money* that the Fed created propped up the banks, but none of them had much appetite for expanding their loan portfolio. The economy contracted, businesses failed and unemployment mushroomed.

Cryptocurrencies have no lender of last resort, and they never will. They live outside the purview of all national or international banking organizations, and they live or die according to their market value. They will never need a lender of last resort. The "money supply" of a cryptocurrency is determined by inviolable rules. Some cryptos are designed to have a fixed supply, others a gradually inflating, but entirely predictable, supply. Their money supply rules are enshrined in software that is guaranteed by fixed rules of governance. This arrangement makes it impossible to print *fractional reserve money*.

This is similar to having a gold standard, where the inviolability of program code, rather than the value of gold, stabilizes the value of the currency. In fact it is more rigorous than having a gold standard for two reasons:

1. With a gold standard, banks can still create *fractional reserve money*. However, if there is a banking crisis, the central bank can only use the reserves it has to stabilize the system. If those reserves are not sufficient then the economy takes the hit. Within a cryptocurrency system there cannot be *fractional reserve money*.

2. The problem with the gold standard was not the standard itself, which did the job it was supposed to do in preserving the value of the currency. The problem was with the governance of the standard. Governments abandoned the standard one by one.

Currently there are rumors of various governments (Japan, Canada, Sweden, Estonia and others) exploring the possibilities of cryptocurrencies. Venezuela has even created one, called the Petro, supposedly backed by petroleum deposits, but there's no evidence of significant use. Russia claims to be working on one, with an interest in using it to circumvent sanctions. But, there is no evidence so far of any government creating a true cryptocurrency, with independent governance.

Governments are attracted to the technological benefits: more efficient payment systems, immutable transaction records on the blockchain and less counterfeiting, but it is difficult to imagine any government deciding to give up control of the fiat currency. So what may happen is that some governments will adopt the technology in a minor way for their own purposes. We cannot see how it can go any further than that.

Neither, at the moment, can we foresee fiat currencies being replaced entirely by one or more successful cryptocurrencies. So, most likely, they will learn to live with each other.

Chapter Summary

This chapter examines some areas of blockchain applications that have already proved themselves to some degree. Here is the list of those covered in this chapter:

- New currencies focussed on inexpensive payment transactions.
- New currencies focussed on micropayments.
- ICOs/crowdfunding.
- Businesses (of all kinds) using tokens as a reward or payment capability.
- Record keeping.
- Video gaming is a natural application for tokens and likely to see phenomenal growth.
- Personal data storage and eventually direct trading of data objects.
- Cryptocurrency banking.

Chapter 9

Customer Acquisition and Permission

Many a small thing has been made large by the right kind of advertising.

~ Mark Twain

―――∞∞∞―――

If customer acquisition and retention were an industry sector it would be massive. In the US alone, the advertising sector garners $220 billion in revenue, about 11.5% of the US work force is employed in sales, and to that you can add millions of web sites, social media staff, SEO, CRM software and the associated computer hardware, and you can include customer support too, and everything that goes with it.

Customer acquisition can be seen through the prism of a sales funnel. The business advertises to attract the interest of potential customers and encourages contact in a variety of ways: through its web site, through dedicated landing pages, via phone, via email, through chat and all of the social media channels. As soon as there is a contact, the prospect has entered the sales funnel.

From then on, how efficiently the business manages the sales funnel is important. Any effort expended on prospects that don't become customers is wasted. So the goal is to qualify prospects in or out as inexpensively as possible. If the sales funnel is well-managed it becomes possible to predict

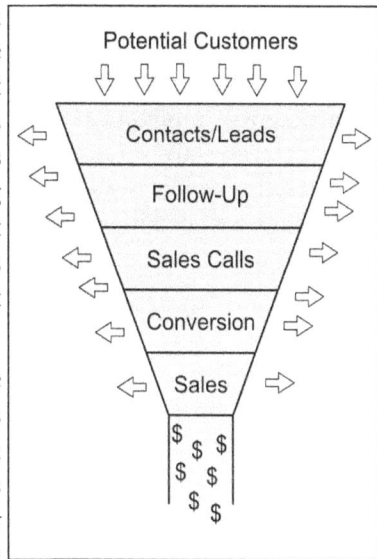

Figure 8. The Sales Funnel

the percentage of prospects who will exit (or be qualified out) at any stage. Consequently it becomes possible to predict how many sales will result if you increase the number of prospects entering the funnel.

Advertising as the Gateway to the Sales Funnel

Customer acquisition is the most challenging area of the sales cycle. For most businesses it is far more expensive to acquire new customers than it is to foster customer loyalty among existing customers. After the first sale, the business may forge a relationship with the customer which can bear fruit for both parties for years. Effective advertising is the gateway to establishing such relationships.

Advertising channels are not organized in such a way as to make it easy for advertisers to engage effectively with potential customers. Prior to digital advertising, all the available channels: TV, radio, billboards, magazines and so on, were broadcast advertising media that were poorly targeted, in the sense that only a small percentage of those who were exposed to an ad had any interest in it. With digital advertising, targeting can be far more effective than that, and it generally is, but it is still far from perfect. Internet advertising never evolved towards constructive engagement between advertiser and prospect.

Interruption Advertising

Perhaps you recall the early days of the Internet, when it was bursting with potential and thousands of websites blossomed into existence. In those halcyon days we never envisaged the digital world that now assails us. For quite a while it was less aggressive than it has recently become. But technology weighed in on the side of the digital ad brokers and the Internet publishers.

It was a natural consequence of the limitations of early internet technologies. There was no practical way to charge a fraction of a cent for reading a webpage, so web publishers sought revenues in the only other way possible, from advertising. With advertising, several cents could be earned by the occasional website visitor clicking on a banner ad or a sponsored link. Within a handful of years ad brokers stepped into the market, acting as middlemen between the advertisers and the web publishers, and before long a real time market in serving ads up like fast food was the order of the day.

The ad brokers co-opted user data so they could present ads to us wherever we went on-line and whatever we did on-line. If, for example, you were interested in buying a laptop and the ad broker got wind of it, you'd be faced with laptop ads even on sites that sold vitamin supplements or published the football scores.

The web experience took a dive. Suddenly news articles were surrounded by intrusive ads and click-bait links. Pop-ups began to pop-up everywhere. Videos you chose to watch were blighted by introductory ads that you never chose to watch. Autoplay videos would start-up unexpectedly. Search results were led by ads masquerading as useful links. The Internet became a virtual blizzard of uncalled for interruption, far more intrusive than ads on the legacy media of TV or radio.

The Internet giants, Amazon, Google, Facebook, Twitter, WhatsApp and many others, joined the party. Some of them entered the ad market with polite and only mildly intrusive business models, particularly the social networks. But their natural commercial motivation to maximize their holy revenues drove them into exploitative relationships with their users. In the domains of these giants, it soon became clear that the advertiser was the customer and that the users had been bamboozled into being the product.

With social networks, it was a natural business goal to entice the user to remain on site for as long as possible, so that as many ads as possible could be presented in as many different ways as possible. From that perspective, Facebook was an almost perfect platform for the interruption-based business model, which is why it became such a valuable company so quickly.

The Dominance of Digital Advertising

Digital advertising now dominates the advertising market and its dominance is increasing. According to a December 2018 report on global advertising market trends from research company MAGNA, global ad revenues grew by 7.2% to $552 billion in 2018, compared to $517 billion in 2017. In the US, growth was 7.5%. By comparison, the US economy grew about 2.9% in 2018.

Digital advertising sales are growing faster than the advertising market as a whole. MAGNA estimated the 2018 growth rate at 17%, giving a total of $251 billion—about 45% of global advertising revenues. MAGNA expects digital advertising to represent half the world's total sales by 2020. Mobile ad sales reached half of total digital spend in 2017 and accounted for 62% in 2018.

Google and Facebook took the lion's share of global digital advertising, with roughly 30% going to Google and 16% to Facebook. In the US, the

2018 figures were 37.1% Google, 20.6% Facebook, according to eMarketer. Amazon has entered the market and, although a distant third, has garnered a 4.1% share.

The Changing of the Guard

The world of digital advertising has several years left before it moves into decline. It is still growing in size and encroaching on the non-digital advertising world, by virtue of its effectiveness. However, beset with waves of bad PR and self-inflicted wounds throughout 2018, Facebook became the "bête noire" of personal data exploitation. It is worth recapping the news stories that have shredded its reputation:

- Before and throughout the 2016 election, it was efficiently exploited by Putin's propaganda machine to spread misleading news reports and flat out lies.

- It happily let its users' data be exploited by Cambridge Analytica, which used it to influence users' voting intention in the 2016 elections. The personal data of over 71 million Facebook users was harvested.

- Facebook is fending off law suits in the US relating to violating its own declared privacy policies and in the EU in respect of GDPR regulation.

- The New York Times revealed that Facebook made deals with 60 device makers (including Apple, Amazon, Microsoft, Blackberry, etc.) allowing them "broad access" to users' data and had been doing such deals since 2007.

- Despite evidence to the contrary, Mark Zuckerberg insisted in his testimony to the US Congress that users had 'complete control' over who could see their Facebook data.

Subsequent news that has since dribbled out reinforces the picture of a business based on the exploitation of its users' data. Facebook's corporate image is now in tatters. It now looks as though Facebook's user growth is starting to stall.

It isn't that Facebook is necessarily the greatest exploiter of user data. It has been suggested, for example, that Palantir, the big data analytics software business, has a more complete and useful set of personal data about US citizens than Facebook has. Other companies, Equifax and

Yahoo for example, have a worse record of keeping user data safe than Facebook. Facebook has simply become the poster child for personal data exploitation.

From the perspective of the user, social networks pursue business models based on interruption advertising, manipulation and deceit. The deceit is fundamental, gulling the user into believing they are receiving a "free service" with the minor irritation of ads and no clue whatsoever about the powerful algorithms and machine learning techniques that will comb through their data to discover opportunities to exploit their behavior. The layout of web pages will be specifically designed to manipulate the user into behavior that generates revenue at the expense of anything else. That's how these business engines work. It has proved to be a lucrative business model.

The User Backlash

Internet users are sick of interruption advertising and they are fighting back. In the US 27.5% (about 75.1 million) have installed browser-based ad blocking on PCs and Macs. Among the Millennials it's higher, estimated to be 41% by eMarketer. The figures are lower for the rest of the world, but the numbers are increasing everywhere. The situation on mobile phones is more complex. It's not possible to block ads with a browser plug-in. And to add insult to injury, your money pays for the ads you never requested. In 2016, Enders Analysis reported that anywhere between 18%-79% of the mobile data transferred from news sites to your phone was ad data, charged to you at your data rate.

The impact of interruption has been studied scientifically. There is a field of science that marries cognitive psychology to the study of human–computer interactions. From the psychological perspective, an interruption distracts you from your activity, causing you to return to it to complete it. Getting back to where you were takes time and energy, and it can take quite a long time depending on how complex the interrupted activity was and at what point it was interrupted. Controlled experiments have demonstrated, indisputably, that interruptions, whether while working in an ER unit, driving a vehicle, or surfing the Internet, reduce performance and cause errors.

Unsolicited ads that interrupt your activity also steal your time. Indeed it can be said that the ad brokers use your data and your computer resources on your dollar to interrupt you and, in doing so, waste your precious time. It's no surprise that many users are doing what they can to prevent this.

The Advertiser Backlash

Advertisers are not happy either. They have different gripes, chief of which is the cost of advertising. Looked at from that side of the market, the cost to advertisers of gaining attention has been rising for years. A Harvard Business School report by Thales S. Teixeira, entitled *The Rising Cost of Consumer Attention: Why You Should Care, and What You Can Do About It,* provides a detailed study of this. In summary, it concludes that the cost of gaining consumer attention (which determines digital advertising ROI) has escalated by a factor of eight over the last two decades, with an approximate growth rate of 10% per year. Over the same twenty year period general price inflation has been less than 2.5% per year, so advertisers are facing 7.5% cost increase year-on-year and they know it.

They know that digital advertising, even if more effective than any other channel, is beset with problems beyond the fact that users resent its interruptive nature. Real time ad placement means that the advertiser has minimal control over the context of where the ad is placed. While their ads will avoid contentious websites, the brokers are targeting the consumer rather than the web page, so ads can appear in places that do not align well with brand image. A particular problem of this kind occurred on YouTube, where ads were appearing alongside offensive content such as jihadist and neo-Nazi videos.

However this is a minor issue when compared with ad fraud, most of which is the activity of scripted software robots (bots) that pervade the Internet. Bots drive fraudulent businesses that make money from automating clicks on ads. The largest such operation, according to White Ops, is Methbot which generates somewhere between $3 to $5 million in fraudulent revenue every day—over $1 billion per year. Pixilate reported that click fraud was running at about 20% in January 2017. And the more the advertiser pays the worse it is likely to be. ANA reported in 2016 that ads with a cost per thousand clicks (CPM) of over $10 experienced a 39% higher level of bot fraud than those with a lower CPM.

Remarkably, there are ad-serving websites that are visited exclusively by bots. According to The Verge, they account for roughly 20 percent of all ad-serving websites. It is rumored that dishonest advertisers participate in bot activity by means of the dark web in order to damage the ad campaigns of their competitors. Ad fraud has spawned a large black market with revenues in the billions.

This parlous state of affairs provides a mild boost to Facebook, because it is less likely that advertisers will encounter bots on Facebook. Yes, there are many false Facebook accounts run by bots—McAfee estimated the number at 161 million—but most are there to spread marketing propaganda, just as Russian Facebook accounts spread political propaganda.

Another problem for the advertiser is the engagement metrics. In most circumstances it is just the click and page view that follows it that is measured. As we indicated earlier, it's the advertisers sales funnel that matters, and particularly the quality of engagement that occurs following some page views. Ad agency Solve conducted an experiment in 2015 to elucidate this. *It managed to generate 100,000 views for a blank video with an advertising spend of just $1400*. It is likely that many of those views were from bots; nobody watching nothing.

On the customer side of the equation we have the fact that distrust of ads is common, if not ubiquitous. The ads are unwanted, as the ad blocking trend demonstrates, and many web users don't pay attention to the ads—preferring instead to research product purchases themselves.

Crypto Currency Ad Businesses

There have been at least 20 blockchain-based advertising business start-ups. There are probably three distinct business opportunities among them:

1. To repair the consumer-publisher-advertiser relationship in a trusted browsing environment.

2. To provide a search-based service that joins Internet users to products after the fashion of Google or Amazon.

3. To create a direct consumer-advertiser relationship in a trusted advertising environment.

Basic Attention Token

Basic Attention Token (BAT) is currently the most successful cryptocurrency business aimed at addressing the first of these opportunities. The company's business model is firmly founded on the fact that it has built and distributed its own web browser, the Brave browser.

The "Common Sense" of Cryptocurrency

This is an amended version of Google's open source Chrome browser, with the distinction that it blocks unsolicited ads and user trackers. It also includes a ledger system that monitors user attention anonymously so that it can reward publishers and users for ad views. In short, it knows how users spend their time, but it does not exploit them.

The underlying cryptocurrency, BAT, pays for services on the BAT platform and is thus exchanged between publishers, advertisers, and users, depending upon the context. Advertisers make payments in BAT and publishers are paid in BAT, with users also capable of receiving BAT payments for their attention to ads, or paying publishers in BAT for the content they consume. Because the BAT platform controls its environment, ad brokers are removed from the equation and the incidence of bots and other forms of fraud are minimized.

The virtue of the BAT approach is that it caters to multiple possibilities: users paying for content, publishers being rewarded by advertising revenue and advertisers rewarding users directly for viewing ads.

Bitclave

BitClave offers a direct advertiser to consumer digital advertising model, based on a search engine. The BitClave environment eliminates all (other) intermediaries. Companies make offers directly to consumers based on their explicit search for goods or services. Consumers control their personal data and can choose whether to reveal their identity or personal information to retailers as part of their search.

Revealing such information may allow retailers to respond to the searches with targeted promotions, which consumers can then be compensated for viewing. The market BitClave creates allows users to use their data to "bid" for products or services they want. In theory this will create a rich ad-based market for consumers.

Permission.io

Note that the author has worked for Permission.io and is at the time of writing retained by the organization as a consultant. He will try not to introduce any bias into the description of Permission.io that follows.

The Permission.io business model is to enable the formation of a direct relationship between consumers and advertisers through ad watching. The business considers itself to be a direct implementation of Seth Godin's ideas on permission marketing: the idea is simply that a far more productive relationship can be created between advertisers and consumers if the interactions between the two are governed by the granting of permission by both parties.

Permission.io has created a "permission marketplace" where advertisers run advertising campaigns and Permission.io users select ads to watch and are rewarded in ASK tokens for doing so. The users perspective is as follows:

- Users' data (profile data) is stored securely and never revealed to anyone except by the express permission of the user.

- Users view ads or other content on the Permission.io web site using a browser. Each ad has an ASK reward attached to it and the user receives the reward if they watch the ad through.

- Users' ASK rewards are stored in a wallet provided by Permission.io.

The advertiser perspective is as follows:

- The marketplace provides an advertising portal for advertisers to build ad campaigns targeting Permission.io users. User data is aggregated anonymously and thus ads are distributed anonymously by matching the targeting criteria to users profiles.

- The advertisers pay Permission.io in dollars for the campaign on the basis of a particular price for a target number of views.

- The advertiser ROI will likely be high because every ad viewer will be a real person (not a bot), only those who are interested will watch and targeting will be precise.

Identity Management

One of the necessities of blockchain based business is to ensure that its users are genuine, with only one identity per person. If possible this needs to be achieved with minimal access to user data. An idea that is being implemented by Permission.io and, I believe, other blockchain businesses is based on the "Wheel of Trust," which is illustrated in *Figure 9* on the next page.

It works like this. Almost everyone who uses the Internet has a fairly large set of declared identity information on social networks, messaging apps, cloud services, eretail sites and so on. If someone wishes to create a fake ID, it is possible to set up fake accounts in a handful of sites. But unless

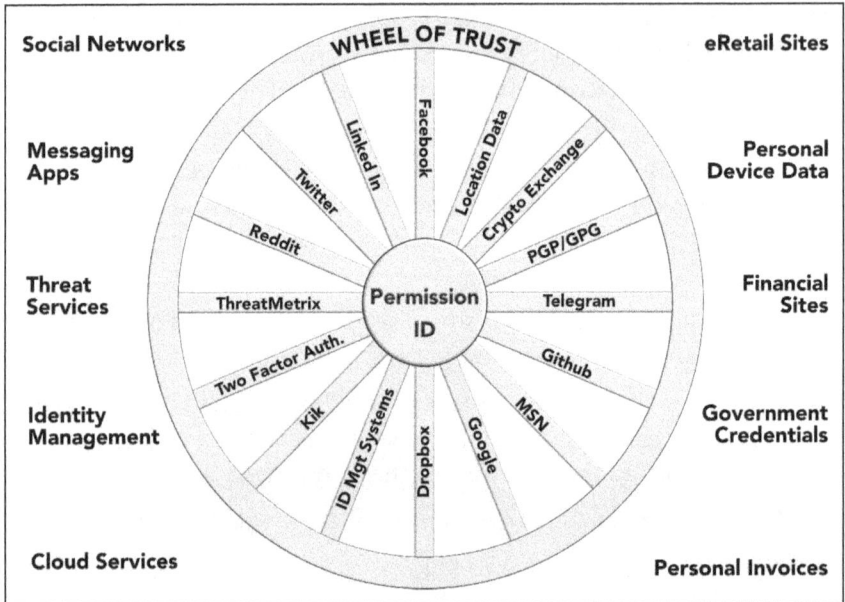

Figure 9. The Wheel of Trust

you are really intent on creating one, you are unlikely to put in the effort of acquiring a large number of fake logins. The effort required to create many convincing fake IDs is prohibitive, unless the potential reward is very large.

In the Wheel of Trust, each set of identity details can be thought of as one spoke in the wheel attesting to the reality of the identity's owner. So users help to validate that they are a real person by willingly providing this data.

Monitoring a large number of such users enables the blockchain company to assign a probability that an ID is fake. To increase the level of certainty, there are identity validation services, such as ThreatMetrix, which can be used. This company, when provided with basic personal details, can attest to the validity of an identity to a very high level of probability for a relatively low one-time fee. Such a service is only one spoke of the wheel, it just happens to be one that offers very high probability of authenticity.

Incidentally, the Wheel of Trust could be offered as a blockchain identity service, and in the long run we expect it to be made available as such. The need to validate identities in a blockchain-based world requires it. In the digital world your digital identity is what you are.

Advertising: Permission, Transparency and Trust

All of the blockchain-based businesses we have discussed are based on data self-sovereignty and permission. A permission-based advertising business is the direct opposite of an interruption-based one. The permission oriented ad model encourages the development of an equal relationship between the prospect and the seller. There is no middleman exploiting user data and harvesting the lion's share of the rewards. The whole arrangement can be transparent in the sense that users, publishers and advertisers can know how the platform works. Bots are banished and click fraud is minimized or eliminated.

We can think of transparency as the opposite of manipulation. Transparency is a natural capability of the blockchain and, in our view, the advertising environments of the future will ride on the back of such technology. If all transactions are recorded and visible to anyone who chooses to audit them, the whole advertising environment will have a different character. In such an environment, trust is firmly established and deceit disappears like a wisp of smoke.

Sadly, the great enterprises of the dot-com era veered off in a distinctly different direction. Their business models are built on interruption, obfuscation, data exploitation, fake bot activity, and a bewildering lack of transparency.

Chapter Summary

This chapter explores the blockchain possibilities in the advertising sector.

– The majority of advertising, and particularly digital advertising, is based on interrupting the user with ads. Digital advertising is now dominated by the Internet ad giants, Google and Facebook, recently joined by Amazon, but also by real time ad brokers. For all such businesses, the advertisers are the customer and the users are the product. This has had the impact of alienating users.

– Digital advertising dominates the advertising market and its dominance is increasing. According to a December 2018 report from research company MAGNA, global ad revenues grew by 7.2% to $552 billion in 2018. Google and Facebook took the lion's share of global digital advertising, with roughly 30% going to Google and 16% to Facebook.

– In the US, 27.5% (about 75.1 million) have installed browser-based ad blocking on PCs and Macs to combat interruptive advertising. The impact of interruption has been studied scientifically. Controlled experiments have demonstrated, indisputably, that interruptions, whether while working in an ER unit, driving a vehicle, or surfing the Internet, reduce performance and cause errors.

– A Harvard Business School report by Thales S. Teixeira, entitled *The Rising Cost of Consumer Attention: Why You Should Care, and What You Can Do About It,* concluded that the cost of gaining consumer attention (which determines digital advertising ROI) has escalated by a factor of eight over the last two decades, with an approximate growth rate of 10% per year.

– We examined two blockchain ad businesses. One, Permission.io, provides a direct consumer to advertiser model, where the consumer is rewarded for watching ads. The other, Basic Attention Token, works with the triangle: consumer-publisher-advertiser, with consumers both able to pay (for content) or be rewarded for watching ads.

– The Wheel of Trust (an idea from Permission.io) offers a powerful and inexpensive way to manage identities.

Chapter 10

Token Sanity

Money often costs too much.

~ Ralph Waldo Emerson

———∞———

If you want to understand cryptocurrencies, it helps to think in terms of existing fiat currencies and then make adjustments to allow for the differences between the two. It helps also to focus on crypto tokens that have a designated use, rather than consider pure crypto coins like Bitcoin, Dash and Monero, whose only function at the moment is payment transactions.

Golem is an example of a crypto token. It presides over a market that enables people or businesses with spare computer power to rent it to people or businesses that need the computer power to run calculation-intensive computer programs. There are hundreds of other such tokens that are designed to enable gambling markets, gaming markets, dental services, advertising, and so on. If a token's primary or only use is in enabling a market, then its value will directly depend on how prosperous that market is—just as the value of the US dollar relates directly to the value of the US economy.

Bubbles and ICOs

The perception of cryptocurrencies has been horribly distorted by speculation, with headlines trumpeting the swings in value of the major cryptocurrencies. This may have been gratifying to those who profited from the speculative excesses of the market, but it only damaged the underlying business of those crypto tokens that had a practical purpose. ICOs, which have thankfully vanished from the US market, were also a distorting factor.

Throughout the ICO mania of 2017, few if any commentators dared to suggest that ICOs were a fundamentally bad idea. And yet it is clearly so. When a token-based cryptocurrency business performs an ICO it sells tokens to investors. Those who buy expect the token to rise in value. The token sold in this way is, by any reasonable definition, a security. An ICO assumes that a token can, without any inherent conflict, be both a security and a currency. In our view it cannot be both. In order to understand why, let us examine the life of a crypto token.

The Token in a Token Economy

The practical value of a token depends on a variety of factors. We illustrate a typical token flow in *Figure 10* on the next page. This shows the various factors that are likely to play a part in a token economy.

To set the scene, we are modeling the situation where a blockchain business enables buyer-seller interactions, and takes a commission on the transactions. The tokens are distributed among a variety of wallets owned by different people or organizations that are involved with this business in some way. We classify them in the following way:

- **Exchange(s).** Holders of the token can buy or sell their tokens on any exchange which makes a market for it. The exchanges determine the market value of the token.

- **Investors.** Most cryptocurrency businesses will have investors. Some may have become investors in an ICO, others may have made direct investments in the token. Once the token appears on an exchange it may attract other investors, who will buy or sell according to their inclinations.

- **Speculators.** We classify speculators as short-term holders of tokens who try to profit from the volatility of the token, rather than view it as a medium to long term investment. (It will become clear why we make this distinction later.)

- **Employees.** Most crypto-businesses give employees bonuses in the cryptocurrency. Eventually they will sell them on an exchange.

- **Stakers/Miners.** Miners (or stakers) who provide resources to run the blockchain are usually paid directly in the cryptocurrency and will, at various times, sell portions of it on an exchange.

- **Agents.** Crypto businesses will normally have some external contractors (developers, recommenders, marketers and so on) who are paid partially in the token. They will periodically sell to an exchange.

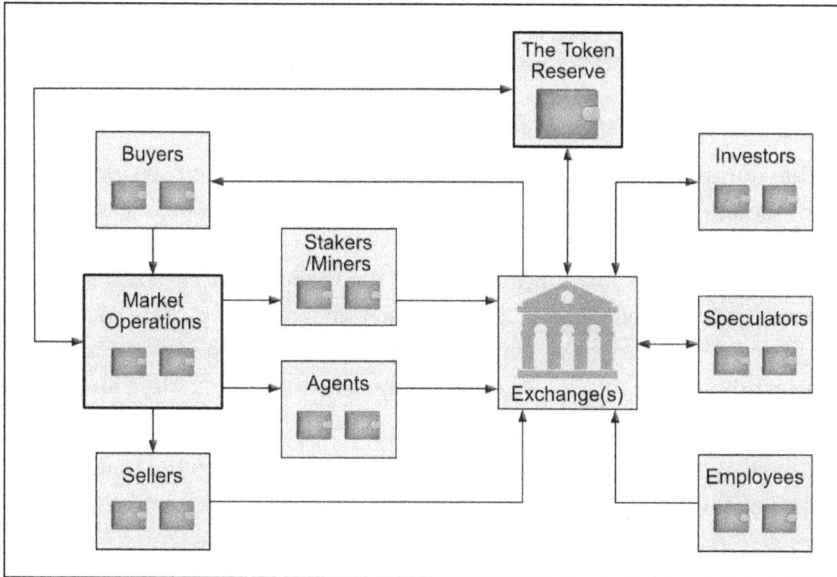

Figure 10. Token Circulation in a Token Economy

- **Buyers and Sellers.** The crypto business enables transactions. The assumption here is that the token is making a market of some kind. For example, the token might be used to make a market for data storage capability. The buyers are storing data using the resources owned by the sellers, who are paid in tokens. The business operating the token economy takes a commission on such transactions. Buyers obtain tokens to pay the sellers, either directly from the crypto business or through an exchange. The sellers ultimately cash in the tokens on an exchange.

- **Market Operations and The Token Reserve.** The crypto business we are modeling holds a fairly large reserve of tokens (The Token Reserve) which, as we shall see, it requires to manage the currency. It will also need to have some tokens in operational wallets, apart from the reserve, to manage market operations: taking commissions, paying miners/stakers and paying agents, and to finance the costs of the business.

There are many actors in this environment who can influence the value of the token. Nevertheless, it is likely that an important goal of the crypto business will be for its token to have a relatively stable price.

The Token Management Goals of the Crypto Business

The two goals stated below are primary goals for a crypto token business.

1. The token value should be relatively stable

The dominant activity of the crypto business in our model is the transactions between buyers and sellers. The crypto business cannot prevent speculators (and possibly investors and staff) from creating volatility in token value, until the use of the token in buy-sell transactions dwarfs the market effects that speculative activity causes. A simple measure of the potential for volatility is the ratio of tokens used in transactions to the total of tokens held by investors, speculators and staff.

The cryptocurrency market is immature, with speculative activity dominating most cryptocurrencies, causing their value to fluctuate wildly. Examine the price graphs of almost any cryptocurrency and you see a pattern of large peaks and deep troughs in price.

This is one of the downsides of a cryptocurrency business. In buying and selling tokens, speculators can be a disturbing influence on the business and both investors and staff can cause volatility when they choose to sell their tokens. If investors or staff have tokens locked up so that they can only be sold after a given date, then there is likely to be a significant downward pressure on the token price on that vesting date. Other holders of the token (buyers, sellers, miners/stakers, agents) are less likely to cause problems.

2. A constant or slowly rising token price is desirable

Cryptocurrencies vary in their approach to the number of tokens in circulation. With Bitcoin, for example, the number of coins increases with every block that is mined, but the rate of inflation of the currency is less than 4% and will gradually decline over time. Others have different inflation rates; some constant and others gradually declining, and some, like Ripple, have no inflation rate at all. Any inflation rate puts a slight downward pressure on price, while a zero inflation rate does not. For this reason we believe that a zero inflation rate is to be preferred.

One of the defining characteristics of a successful currency is that it acts as a metric of value, enabling people to compare prices and estimate what offers the best value. Currently, no cryptocurrencies are anywhere near achieving that, even for a small number of their regular users.

Where the crypto business is in direct competition with a non-crypto business, buyers and sellers will compare prices between the two businesses, and they will use the fiat currency they are familiar with to make their comparisons. The crypto business needs to accommodate this. It needs to use fiat currency for pricing, even if the token's value is relatively stable. The natural solution to this is for the crypto businesses to set prices in the local currency (Dollars in the US, Yen in Japan and so on) for their customers and simultaneously display the equivalent token price. The only exceptions we can think of are cryptocurrencies like Augur, which are pursuing entirely new business models.

As a rule of thumb, buyers and sellers need to be insulated from fluctuations in the cryptocurrency and the only way to achieve that is for the cryptocurrency to be stable or steadily increasing in value. The key to stability in the token price is for token use within the token economy to increase, without its price increasing too quickly.

Incidentally, this also applies to pure cryptocurrencies, like Bitcoin, Dash and Zcash. There is no possibility of any of them replacing any stable fiat currency unless their volatility decreases dramatically and they can be trusted to maintain their value.

Managing the Token's Actual Value and Market Value

The actual value of a token in use and its market value, as indicated by its price on exchanges, can be very different. The two prices are driven by entirely different influences. The performance of the token economy drives the actual value of the token; investment enthusiasm drives the market value.

A token economy, as illustrated by the model in *Figure 10*, is a semi-closed system, built to generate an operational profit. The activity of the system can be measured using the token, but it can also be measured using dollars (or any other appropriate fiat currency) as long as its transactions are also priced in dollars. It is possible to measure the number of tokens that circulate over any given time period by examining the blockchain, if the dollar equivalent prices are also stored there. If we consider the token circulation over a year in terms of dollars, this is the amount buyers paid in a year, say x. The equivalent in tokens can also be known, by taking the token price of each transaction to arrive at a different total. Let us call this y.

Thus if we make this calculation over a year, we get the simple equivalence that:

$$\mathbf{y} \text{ tokens} = \$\mathbf{x} \text{ , and}$$

$$1 \text{ token} = \$\mathbf{x}/\mathbf{y}$$

We can make the same calculation for any time period, a month, a day, an hour and so on. As the token's price fluctuates against the dollar, a longer term figure will provide a better estimate of value in any given time period. Simple mathematics can identify trends in value and it is reasonably easy to construct a time-weighted average to give a fairly accurate value of the token at any given time. In a nutshell, the crypto business can estimate with reasonable accuracy what the dollar value of its token actually is.

Because the crypto business can know this, and know it faster than any other holder of the token, it can take advantage of this knowledge. It can sell its tokens on an exchange when the price rises high above the actual value of the token, or buy tokens if the price falls far below the actual value. This might be thought of as price manipulation, but if the crypto business publishes its price estimate in real-time, then it isn't, because the market is as well informed as the business.

The crypto business buying and selling its own tokens can be thought of as a treasury function. It is the same activity that central banks indulge in when trying to defend the value of their national currency. Central banks hold reserves in precious metals (gold and sometimes silver) and also in the major trading currencies. They do not want the national currency to fluctuate too greatly, because it has a direct impact on the cost of imports and the price of exports. So they try to keep it within a fairly tight trading range.

If the balance of trade eventually causes a drain on their reserves, they allow the currency to devalue against the major trading currencies. If the currency's exchange rate gradually rises against other trading currencies, they increase their reserves to put a downward pressure on the exchange rate. If the trend continues, when they consider that they have sufficient reserves, they allow the exchange rate to rise.

So the crypto business could behave in a similar way and could even pre-announce the points at which it intends to intervene, as central banks sometimes do.

The Crypto Business Treasury Duties

Crypto businesses need to manage their currencies in a similar way to fiat currencies. To do so, they need to consider the "profitability" of their crypto economy. Returning to our simple model, most of a crypto business's costs are the same as for any other business: infrastructure costs, operational costs, staff salaries and so on. These have to be covered by the commissions earned on transactions.

If we represent the total of all sales within the crypto economy over a year as **y** tokens, the business costs as **c** tokens and the realized profit as **p** tokens, then we can get the simple equivalence:

$$y = c + p$$

The first thing to note is that if there is a loss rather than a profit (i.e. if **p** is negative) the crypto business will need to fund its normal business operations from its Token Reserve. If the situation persists this will eventually destroy the business as the reserve disappears. A crypto business is more exposed to the impact of losses than a conventional business, because its business activities are visible in real time on the blockchain. A private business is not obliged to report much. A publicly quoted business only needs to report its financial situation every quarter (and those quarterly reports can be misleading). A crypto business has no easy means to obfuscate its financial situation.

A crypto business will have some expenses that need to be paid in the local fiat currency. It will thus need to budget for them and to hold an appropriate amount of fait currencies to meet them. On the one hand it may not want to make frequent transfers to a fiat currency account (it is likely that the transaction cost will be lower if fewer such transfers are made), and on the other it may not want to hold too much fiat if its token is rising in value against the fiat currency. Managing this treasury function is a matter of mathematics: plotting trends and using them to make least-cost decisions.

Putting such expenses to one side, the crypto business needs to choose its own "metric of value." Crypto businesses are naturally global, so it may be a mistake to take the local fiat currency as a measure of value of the token, unless the majority of the crypto business's users are concentrated in that currency. Depending on the distribution of that user population, it could choose instead to measure its token value against a basket of fiat currencies.

The situation could be more complicated than that. It may be that many of its users are also holding another crypto currency, like Bitcoin or Ether, and frequently transfer their tokens into those crypto currencies. Thus it may need to include those crypto currencies in its basket of currencies.

To add a further complication, it could be the case that a general inflation is taking place in the world of fiat currencies and thus it needs to take account of this too. The value of its token may appear to be rising, but this may only reflect fiat currency inflation, rather than a true increase in value. To counter such possibilities, the crypto business can choose to do what national banks do and hold reserves of gold or other precious metals to offset currency inflation.

Putting this all together, it is clear that there is need for a treasury function within a successful crypto business, which involves holding reserves in a similar way to how national banks hold reserves of other currencies and precious metals.

Tactical Aspects of Token Management

As we have noted, the goal for the crypto business will most likely be for the token to gradually rise in value in relation to other currencies and particularly in relation to any competitive currency, whether fiat or otherwise. The crypto business will not be able to ensure this unless its activities are profitable. If they are, then the profit will manifest as an increase in its share of the whole token supply. It should then be able to put its profits to use.

The question is: how can it use its increasing supply of tokens to raise the value of the token in a relatively stable way?

Before we focus on possible tactics, we need to discuss currency circulation. We need to consider a simple money supply equation, which measures the velocity of money within an economy.

It is usually written as:

$$MV = PY$$

where M is the supply of money in an economy

V is the velocity of money

P is the overall price level, and

Y is the gross domestic product (GDP) of the economy.

On left hand side of the equation, we have the movement of money caused by various businesses or individuals buying things. If the money supply were, say, $100,000,000, and if all the purchases made within a year amounted to $200,000,000, then the money supply would have (on average) circulated twice in the year.

On the right hand side of the equation we have price levels and the GDP. This is the total of every item sold in the economy and the price at which it was sold. So this is not a complicated equation. All it is saying is that if you take everything sold and the price at which it was sold and you know the money supply, then you can work out the average speed of the money supply.

This simple equation is interesting, because it provides an easy way to deduce the relationship between money and the sale of goods and services. If, for example, the GDP increased and prices remained the same, then the supply of money would need to be increased, or the money would have to move faster. It is possible for money to move faster. If you can spend money faster than you usually do, and many other people do too, then the velocity of money will rise.

However, if that doesn't happen when the GDP increases, the increase in GDP will put a downward pressure on price; either there will be lower prices or some of the GDP will simply not be sold.

Now consider the situation where prices increase but the GDP does not. In that situation, either the velocity of money went up or the supply of money increased.

You can also ask: what happens if the money supply increases, but the GDP decreases? In that situation the velocity of money stays the same, and the prices rise.

All of that can be deduced from the $MV = PY$ equation.

Having digested that, we can now apply the same kind of equation and the same logic to a crypto currency ecosystem.

$$MV = \sum PS$$

where M and V are the money supply in tokens and its velocity as before, and the other side of the equation is the sum of all sales, S, and the prices, P (in tokens), at which the sale was made.

We can now use this equation to discuss the tactics a crypto business can use to push up the price of its token within its ecosystem. They are as follows:

1. Increase the number of buyers and sellers

It is should be no surprise that an increase in sales within the ecosystem will push up the price of the token. If there are additional transactions then the crypto business will receive more tokens and its profit (after costs) will increase its holding of tokens. If it retains those tokens then it will reduce the money supply, which will, in turn, put an upward pressure on the token price (assuming the velocity of the tokens does not increase).

2. Diversify by introducing new types of transactions

This is not much different to simply increasing the numbers of buyers and sellers. Just as Amazon expanded from books to CDs and DVDs and then gradually diversified to many other things, the crypto business will prosper through diversification and it will raise the token price. It is important for the crypto business to enable buyers to also become sellers and sellers to become buyers. To do this it may need to diversify considerably.

In reality almost everyone is both a buyer and a seller in their personal activity. If they are employed they are selling their labor (if nothing else), and they buy things with what they earn. Businesses are the same, they have outputs (sales) and inputs (costs).

A national economy is well-balanced, as all participants are inevitably buyers and sellers. A crypto business is unlikely to be well-balanced in that way. Consequently, it will have little control over the entry of tokens (from exchanges to buyers) into its ecosystem and their exit from the ecosystem (from sellers to exchanges). A balance of buying and selling within the token ecosystem will serve to keep tokens within the ecosystem.

Incidentally, because of the issue of balance, in the future we can expect there to be mergers between crypto businesses for the sake of expanding compatible ecosystems.

3. Act as a cryptocurrency exchange

We are not suggesting that crypto businesses should try to compete with large cryptocurrency exchanges, trading in many different tokens—only that they should make a market in their own tokens and the other currencies that they need to interact with (fiat currencies like the dollar and possibly other prevalent cryptocurrencies like Bitcoin and Ether). The obvious reason to do this is to make it easy for holders of the token to buy and sell it with fiat currency. Additionally, if the business can charge lower fees than the exchanges and still be profitable, this activity will support the price of the token.

4. Reduce the token supply

The supply of tokens affects the token price. If there is a fixed supply (a zero inflation rate) it puts a natural upward pressure on the token price. Tokens with a gradually inflating supply can also put an upward pressure on the price if the fiat currency in which the token is valued is inflating at a higher rate. However, the situation is likely to be a little more complicated than that.

If the crypto business holds a significant reserve of the tokens, it can use them to increase or decrease the number of tokens in circulation. If the circulation is increased, it will put a downward pressure on price, and if decreased, it will create an upward pressure. Simply by making profits and adding them to the reserve, the crypto business will put an upward pressure on the token price.

It is possible to incentivize users to hold the token rather than exchange it and this will reduce the token circulation. This can be done, for example, by paying an interest rate to those who hold a designated amount of tokens for an agreed time period. Determining the appropriate interest rate will involve mathematical modeling. The arrangement can be presided over automatically using a smart contract.

5. Reduce the token velocity

Like reducing the token supply, reducing the velocity of tokens pushes up the token price to some degree, and increasing it pushes it down. The crypto business needs to be aware of this and needs to monitor the token velocity. It will discover that some token holders turn over their holdings rapidly and others do not.

It's common knowledge, for example, that about 30 percent of people with air miles never use them. This effect will be the same with token holders in most crypto businesses. Those holding small amounts that are growing slowly, if at all, will probably leave them where they are. Unless its token holders complain, the crypto business has no motivation to increase token velocity. Depending upon its business model it may be able to charge a premium for fast processing. Bear in mind that it is likely to be faster to clear token transactions than for banks to clear fiat transactions, which can, under some circumstances, take a day or more.

6. Encourage token users to become "stakers"

Most crypto businesses have stake holders (or miners) who provide computer resources to create new blocks on the blockchain. They are paid in tokens for the service and usually participate in the governance of the blockchain. A crypto business can create a scheme where a token holder (or a group of token holders) assigns a number of tokens to funding the set up and operation of a staking node. The staked tokens then begin to provide the token holder with a steady income of tokens through "staking."

This activity removes tokens from circulation and thus puts an upward pressure on the price of the token. If managed well, this also benefits blockchain governance, as it contributes to decentralizing the blockchain network and increases the number of genuine stake holders who have an interest in the success of the network.

7. Ensure that the token is not volatile and that its value rises

The primary goal in token management is to achieve a non-volatile token price that gradually rises. We already described a number of token management tactics that can help, by putting an upward pressure on the price. However, beyond that there is an overall objective that a crypto business could and perhaps should have. It is this:

To encourage token holders use the token as money.

Consider these three situations:

1. Token holders only use the token for the services the crypto business provides.
2. Token holders hold tokens as an investment.
3. Token holders use their token wallet as a checking account.

In the first situation, the token holder maintains a balance of tokens in their wallet to use the services of the crypto business. Most likely that will be a small proportion of their regular income. In the second situation the token holder holds a significant amount of tokens in their wallet as an investment. As such it is similar to a savings account. In the third situation the token holder holds a fraction of their regular income in tokens in the wallet.

No crypto business will tempt its users to use their wallets as savings accounts unless the token price can be seen to be rising (even if there are dips in value from time to time). And they will not use their wallets as a checking account (in preference to fiat money) unless the token price exhibits very little volatility and gradually rises.

All tokens can be investments and genuine currencies, but to achieve that requires a token that is gradually appreciating and is non-volatile. And if that is achieved to any degree, then the demand for the token will rise under its own force.

Chapter Summary

Chapter 10 discusses the important treasury function within a cryptocurrency business.

- The primary token management goals of a crypto business will most likely aim at minimizing the volatility of the token value. Hence they are likely to be:

  > To keep the token value relatively stable.

  > To encourage a constant or slowly rising token price.

- A crypto business can know (or estimate) the value of its token far more accurately than any other agency and, in particular, far more accurately than a cryptocurrency exchange. For this reason, a cryptobusiness could stabilize its token by declaring its realtime value, and automatically and transparently buying or selling when the value fell unreasonably low or rose unreasonably high.

- Aspects of token management were discussed, suggesting the following as sensible token management tactics.

  > Increasing the number of buyers and sellers in the crypto economy will tend to raise the value and hence the price of the token.

  > Diversifying by introducing new types of transactions will tend to raise the value and hence the price of the token.

  > Making a market in one's own tokens with fiat currencies will likely help to raise the value and hence the price of the token. (But this is a double edged sword, the easier it is for value to enter the economy, the easier it is for value to leave.)

  > Any action that reduces the circulating supply of tokens will help to raise the value and hence the price of the token. Encouraging token users to hold rather than transact has this effect.

  > Any action that reduces the speed of circulation of tokens will help to raise the value and hence the price of the token.

  > Lower price volatility will help to raise the value and hence the price of the token.

Chapter 11

The Footfall of The First Revolution

They that can give up essential liberty to obtain a little temporary safety deserve neither liberty nor safety.

~ Benjamin Franklin

———— ◦◦◦ ————

Some inventions are world-changing. Here's a list of some important ones: the nail, concrete, the stirrup, the compass, the cannon, the mechanical clock, the windmill, the light bulb and the camera. Select an item from the list and research its impact on mankind. It will tell a tale of dramatic change wrought by a single practical idea. But search as you may, you will not find anything prior to the 20th century that had anything even close to the impact of Johannes Gutenberg's printing press.

To be clear, Gutenberg did not invent printing itself. Woodblock printing was invented in China in the 7th century AD and had migrated to Europe by Gutenberg's time. What Gutenberg invented was distinctly different technology. It was a metal moveable-type printing system that could be automated in a production-line fashion. It accelerated the printing process substantially and tore down the cost.

Naturally, Gutenberg chose The Bible for his first publishing project. Before the printing press gradually consigned them to irrelevance, there were whole monasteries full of monks devotedly copying out The Bible by hand—devoted Xerox machines made possible by bands of devoted brothers and their devoted sponsors.

The Gutenberg Bible was available only by pre-order and records suggest that fewer than 200 copies were printed over a period of four years. A copy could be bought for a price equal to about three years' wages for the average clerk. Publishing one Bible per week may not seem particularly productive until you compare it to the competition. The increase in productivity was dramatic and the value of a printing press was quickly recognized. So Gutenberg's technology infected the rest of Europe, making its way to Cologne, Amsterdam, Rome, Venice, Paris, and London.

Revolutionary Vocabulary

William Caxton, who brought the printing press to London, first encountered it in Cologne. He set up a printing press in Bruges, in 1473, where he printed his first English book, Recuyell of the Historyes of Troye. By 1476, Caxton was in London, printing The Canterbury Tales, by Chaucer. Almost immediately, however, Caxton's printing business encountered a problem: at the time there were only five books written in English. Caxton needed content, so he decided to print translations of books in French, Latin, and even ancient Greek. Translators were in short supply, so he did the work himself—with a distinct lack of scholarly accuracy—and in the process he provoked a profound metamorphosis of the English language. The language shifted from what academics refer to as Middle English to Early Modern English.

Dictionary.com claims that about 80 percent of the words in any English dictionary are borrowed, mainly from Latin, and 10 percent come directly from Latin. Yet, the Middle English of Caxton's era had almost no Latin words—having evolved from the Anglo-Saxon language. The Latin words, and those with Latin roots. made their way into the language by way of publishers' translations of Latin and French works.

The printing press provoked a metamorphosis of the English language that would soon see the playwrights of the Elizabethan era experimenting with new literary forms and inventing hundreds of new words.

Revolutionary Bibles

The Gutenberg Bible was a reproduction of the Vulgate Bible, the Catholic Church's official Latin text that dated back to the 4th century. In the wake of the printing press the idea soon emerged that The Bible could be translated into other languages. In 1516, in the preface to his translation of the New Testament from Greek into Latin, the Dutch scholar, Desiderius Erasmus, stated his view that the holy book should be available in every language. Literacy was on the rise, and the opinion that scripture was what ordinary people needed for their spiritual welfare was difficult to oppose.

The idea became a central demand of the Protestant Reformation that began in 1517, led by Martin Luther, with his publication of the famous Ninety-five Theses. In the Theses, Luther criticized both the wealth and the behavior of the Catholic Church. The Church tried to suppress Luther, but proved incapable either of arresting him or silencing him. Working in Wartburg under the protection of Frederick III, the Elector of Saxony, Luther had a German translation of the New Testament ready for publication by 1522. His translation of the Old Testament appeared later, in 1534.

The monopoly the Catholic Church held over the publication of The Bible and its interpretation was broken and its power diminished as new translations of the Bible proliferated. In the years that followed, most of Northern Europe became Protestant, in one form or another: Lutheran, Calvinist, Anabaptist and, in England, Anglican. The schisms led to civil strife and war throughout the 16th and 17th centuries, the so called "Wars of the Reformation."

In the 17th century there was a large wave of immigration into the British North American colonies. The new arrivals were religious refugees escaping persecution in Europe. The descendants of these Protestant settlers would later invent a totally new country.

Revolutionary Science

Galileo's dispute with the Catholic Church—whether the Sun revolves around the Earth, or the Earth around the Sun—was pivotal in establishing the European universities, and later, scientific societies, as the definitive source of knowledge. From Galileo's side the mechanics of the solar system was a scientific question; from the Church's side it was a religious question. The Church's opinion rested on a literal reading of several parts of the Bible, including Psalms 93, 96 and 104, all of which suggest that the Earth does not move.

The difference of opinion turned into dangerous conflict in 1615 when Galileo's heliocentric theory was brought to the attention of the Catholic Inquisition, and it culminated in 1633, with Galileo's trial and condemnation. He was sentenced to prison, but the judgement was commuted to house arrest, under which he lived out the remainder of his life. At the time the Catholic Church maintained an Index of Forbidden Books (Index Librorem Prohibitorum) and books that advocated the heliocentric view of Galileo and Copernicus were added to that list. Following the Galileo controversy, the Catholic Church rarely ventured into scientific debate, except in situations such as Darwin's Theory of Evolution, where the proposed scientific theory flatly contradicted the Church's interpretation of its sacred texts.

To its credit, throughout most of its history, the Catholic Church invested in and promoted education. Some of the early universities evolved from cathedral schools or monastic schools, while others were established by monarchs. In the Middle Ages, much of Europe's education infrastructure was overseen by the Church. However, its influence on the curriculum rarely extended beyond religious education; it opposed scientific ideas only when scientific ideas opposed The Bible. So the universities evolved as self-governing institutions of scholars, with the freedom to speculate in a wide range of fields.

However estimates suggest that only 5 percent of medieval Europe received any kind of schooling; the universities were attended only by the sons of the wealthy and the aristocracy. Literacy was a luxury and illiteracy was the norm for the bulk of the population. But with the invention and proliferation of the printing press, literacy slowly spread (see Table 3).

The level of literacy, which had been fairly constant throughout the Middle Ages, began to rise and then took off for the sky, passing the 50% mark around the beginning of the 19th century. This increase ran parallel to the rise of science. An illiterate population that had passively received its beliefs and world view from the Church, began to yield to the influence of science, and publishing naturally played a role in the shift. The growth of literacy went hand in hand with the development of an intelligentsia that had a distinctly different character to the Jesuit intelligentsia of the Catholic Church.

Literacy Levels in Europe	
Date	Percentage
Middle Ages	5.0%
1475	9.8%
1550	16.2%
1650	37.8%
1750	45.8%
1820	50.2%
1870	75.0%
Data from **Our World in Data**	

Table 3. Literacy Levels

Historians sometimes refer to the span of time from Copernicus and Galileo to the late seventeenth century as the Scientific Revolution. Because Copernicus never published his work until shortly before his death, Galileo is usually credited with kicking it off, not just for the stand he took against the Catholic Church, but also for his astronomical discoveries and his experiments in mechanics. Nevertheless, the philosophical underpinnings of the Scientific revolution were first proposed long after his death, by Francis Bacon (1561-1626), "the father of empiricism." He defined the scientific method.

He struck a note that resonated widely and wildly when he wrote that "knowledge and human power are synonymous."

The Scientific revolution is usually deemed to have culminated with Sir Isaac Newton, a brilliant mathematician and an unparalleled physicist. He single-handedly invented the physics of light and defined the model of the astrophysical universe that held sway for 200 years—until Einstein offered his humble alternative. Newton was a polymath, an author, an alchemist, and a theologian. He was the president of the Royal Society, a Member of Parliament for several years, and Master of the Royal Mint. It was in that role, as we've mentioned, that he defined and implemented the gold standard for the British Pound.

In praise of Newton's extraordinary contribution to science, Alexander Pope wrote the immortal words:

> Nature and Nature's laws lay hid in night:
> God said, Let Newton be! and all was light.

Science was hailed as the "triumph of reason," and Newton was its patron saint. And it was not long before "reason" infected disciplines far outside the orbit of science.

Revolutionary Reason

The Enlightenment, or The Age of Reason, as it is also called, rode on the back of scientific thinking. If the nature of the universe, from the atom to the galaxy, could be examined with the tools of reason, then so could the affairs of man. This opened the gate to new thinking in philosophy, in economics, in politics, in art and in literature. The ideas of Francis Bacon were embraced and extended by Renee Descartes, John Locke and David Hume. In an influential essay entitled, *What Is Enlightenment?*, Immanuel Kant gave The Age of Reason a motto:

Dare to Know! Have courage to use your reason.

Enlightenment thinkers throughout Europe started to question traditional authority. They dared to suggest that all the institutions of humanity could be improved through rational change. Reason, they insisted, was the primary source of legitimacy and authority. Rational discourse could resolve most if not all disagreements. They discussed and debated the rights of man, and they promoted newly minted ideals and ideas: liberty, progress, equality, tolerance, government by constitution and the separation of church and state.

153

They printed books and pamphlets, broadcasting their ideas through the mass medium of that age, the printing press. Reason undermined the authority of kings and pulled the rug from under religion. Reason sailed across the Atlantic Ocean to the British colonies. On arrival, it took root among the intelligentsia of the colonies and inculcated them with European ideas.

The Age of Reason spawned two revolutions: one in America to the west of the Atlantic Ocean and the other to its east, in France.

The American Revolution

The American Revolution was distinctly unusual. The Revolutionary War was a war, not a bloody revolution in the style of the French Revolution or the Russian Revolution. The American casualties were soldiers, even though they were volunteers. No more than 48,000 men served in the Continental Army at any time, although 231,000 men served in total. The American casualties are estimated at about 6,800, with a further 17,000 dying of disease. The British casualties were professional soldiers too, numbering about 25,000, with about two thirds falling to disease.

The British army included about 25,000 loyalists who made themselves scarce when the British lost. Indeed around 80,000, from a population of 2.5 million Americans, emigrated to Canada, or back to Britain, to avoid recriminations. There were incidences of loyalists' property being looted and burned, but there were no public executions or massacres. The Founding Fathers, honoring their ideals, chose to establish a republic, created a written constitution, and enshrined within it a Bill of Rights.

Politically and philosophically, the founding of America was the peak achievement of The Enlightenment. As history demonstrated, the colonies were fertile ground for such a development. The populace was relatively well-educated and receptive to new political thought. They resented the high-handed dictates of the distant British monarchy and its government, regarding it as an infringement of their liberty. There was no entrenched aristocracy to displace. Freedom of religion was a de facto reality, established by the early settlers. Hence there was no resistance to the separation of church and state that the First Amendment guaranteed, and neither was anyone likely to challenge its insistence on freedom of speech and freedom of the press. Such freedoms had spawned the American Revolution and were required to preserve its future.

With the benefit of hindsight, it is easy to criticize the US Constitution. Its original concept of democracy was imperfect. Seven of the seventeen amendments later added to it were safeguards of civil rights. It took a bloody civil war and the remarkable leadership of Abraham Lincoln to abolish slavery and remind America of its original ideals. And it wasn't until 1920, when the 19th amendment (women's voting rights) was ratified, that universal suffrage was achieved,

The Territorial Monopolies

A primary consequence of The Enlightenment was the destruction of powerful political monopolies. The British monarchy lost one of its valuable colonies, and as a consequence extended a greater degree of self-governance to its Canadian colonies. Nevertheless, British Imperialism was undiminished. In the 18th and 19th century, Britain expanded its territories across the world, taking control of a large swathe of Africa and the Indian subcontinent, as well as parts of South East Asia, Australasia and a variety of strategic islands for the sake of its powerful navy. Other European nations —France, Italy, The Netherlands—and Russia also took part in a global territorial land grab.

America, too, was expanding. With the Louisiana Purchase, in 1803, the US acquired a huge area of land. Spain ceded Florida to the US in 1819. Texas was annexed in 1845, Oregon and Washington in 1846, and the other Western states were added as a result of the Mexican-American war. The United States was distinctly different in how it treated these territorial gains, preferring to assimilate these new territories as states within its federal system. Their populations became American citizens.

By the end of the Second World War, the European Empires were in steep decline, a decline that culminated in 1989 with the collapse of the Soviet Union. A whole host of previously Communist countries opted for democracy and independence from the Soviet Block. Many of them, including newly spawned countries that emerged from the collapse of Yugoslavia, threw in their lot with the European Union.

Politically, the centuries following the Revolutionary War could be viewed as a contest between the forces of democracy and the forces of authoritarian rule. If so, the contest is not entirely over, but it is clear who is winning. Currently, about two thirds of the nations of the world are democracies, although some would better be described as flawed democracies. The remainder, 52 in total, have authoritarian governments.

The Economic Landscape

Economics and politics are inextricably related. Maslow's Hierarchy of Needs tells the story. It is illustrated below in Figure 11.

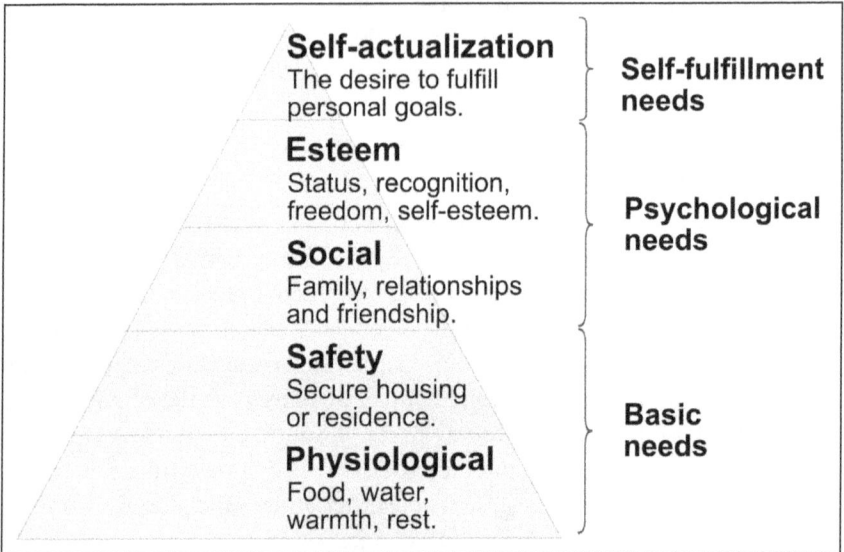

Self-actualization
The desire to fulfill personal goals.

} **Self-fulfillment needs**

Esteem
Status, recognition, freedom, self-esteem.

Social
Family, relationships and friendship.

} **Psychological needs**

Safety
Secure housing or residence.

Physiological
Food, water, warmth, rest.

} **Basic needs**

Figure 11. Maslow's Hierarchy of Needs

Basic human needs have two aspects: bodily necessities (food, water, etc.) which take precedence over Safety (secure housing or residence). Psychological needs occupy the next higher level in the hierarchy. Here Social needs (family, relationships, etc.) take precedence over Esteem (status, recognition, freedom and self-esteem). At the top sits Self-actualization, the desire to fulfill personal goals.

No economic system can support the higher needs in Maslow's pyramid if it fails to satisfy the Basic needs. Authoritarian regimes are instinctively aware of this and subjugate their populations, either by forcing most of the population to be fully focused on their Basic needs or by cultivating dependency, so that Psychological needs and Self-fulfillment needs can only be achieved by obedient conformity to the political system. In such an environment, corrupt behavior and selective patronage may even be deliberate government policy.

Non-authoritarian regimes do the opposite. They cultivate and, at times, champion the individual goal of freedom. A famous and enduring statement of this, penned by Thomas Jefferson, forms the second paragraph of the United States Declaration of Independence. It reads:

> *We hold these truths to be self-evident, that all men are created equal, that they are endowed by their Creator with certain unalienable Rights, that among these are Life, Liberty and the Pursuit of Happiness. That to secure these rights, Governments are instituted among Men, deriving their just powers from the consent of the governed.*

You may see this as idealistic, but it aligns naturally with Maslow's Hierarchy of Needs. And while, at various times throughout history, the United States may have been less than perfect in its pursuit of its founders' ideals, they are nonetheless enshrined in The Declaration of Independence.

If we look at a modern, developed economy from the perspective of Maslow's Hierarchy, it becomes clear that only a relatively small percentage of commercial activity is devoted to people's Basic needs—perhaps as little as 10 - 15%, depending on how you choose to count it. It is nowhere near even 50%. And yet a few hundred years ago it was close to 100%.

Economic Revolution

As for the US economy, it is difficult not to regard it as a revolutionary development. One cannot help but see the visible hand of Adam Smith, The Enlightenment's premier economist, as having guided it forward. At the time of the Revolutionary War, the American economy was dominated by agriculture, and supplemented by cottage industries and a limited amount of shipbuilding for trading purposes. Industrialization only began in the 1790s and early 1800s, with textile manufacturing in New England. It wasn't until a decade after the Civil War that the US economy quickened, entering a long period of rapid growth, stimulated by railroads, mining and a growing manufacturing sector that could now transport its products across a thriving nation.

In the mid-19th century, American wages, especially for skilled workers, were significantly higher than in Europe. That disparity, coupled with the fact that the US was sparsely populated and welcomed immigration, led to an influx of millions of European immigrants. This increased the labor force, which in turn boosted GDP. It also spurred development of the Western states, mainly in farming, ranching and mining. And even as the wage disparity with Europe leveled off, the waves of immigration continued.

Those propitious economic tail winds do not, on their own, explain how America became such a remarkable engine of economic growth. The primary cause was, in our view, innovation in the financial sector.

It began with Alexander Hamilton, who became the US Secretary of the Treasury in 1789 and architect of the nation's financial policy. With the approval of Congress, he chartered a Bank of the United States, as a partly publicly and privately owned corporation. Private investors could purchase stock by tendering the new federal debt that had been issued. Within 2 years the national bank was fully subscribed.

An important consequence of this was that a US banking industry began to flower. By 1810 there were 101 new state-chartered banks; by 1840 there were 834. The capital of these banks increased from $3 million in 1790 to $426 million by 1840. These state-chartered banks were limited liability corporations that attracted a great deal of capital investment, both within the US and from abroad. America had given birth to the first competitive banking market, when banking in the rest of the world was structured around guaranteed monopolies. At that time London was far and away the world's leading financial center and the pound sterling the dominant currency, but by 1825 American banks had about 2.4 times the banking capital of the UK.

The growth of US banking was paralleled, soon after 1790, by the emergence and growth in debt and equity markets, in New York, Philadelphia, Boston and Baltimore. There was trade in U.S. debt issues and Bank stock, which soon expanded to include local equity and debt securities. Remarkably, by 1803, over half U.S. government debt and the stock of the Bank of the United States and all American securities issued were held by European investors. By 1825 there were a total of 232 securities listed on US exchanges, compared to 320 securities in London. The US securities markets had become prominent in a very short period of time.

It could thus be argued that the US was a major commercial power long before the Civil War started. As the decades rolled forward it became the exemplar of a lightly regulated capitalist economy and it soon came to be seen as the most innovative nation in the world.

Chapter Summary

This chapter describes the revolutions that were spawned by Gutenberg's printing press. They were:

– Revolutionary changes in language (English offering an excellent case in point) as societies moved from the spoken word to the written word.

– Revolutionary changes in religion when the Catholic Church lost its monopoly on publishing The Bible.

– Copernicus and Galileo revolt against the biblically based doctrines of the Catholic Church, challenging its authority.

– Revolutionary increases in literacy across all advanced economies, widening the market for books, magazines and newspapers.

– The Age of Enlightenment/The Age of Reason: the intellectual and philosophical movement that dominated 18th Century Europe and the Colonies.

– The political revolution in America, France and elsewhere, destroying monarchies in favor of republics and championing democracy.

– The economic revolution, provoked to some degree by the influence of Adam Smith, resulting in the triumph of capitalism, primarily in America but spreading far and wide.

The "Common Sense" of Cryptocurrency

Chapter 12

The Path of the Second Revolution

We have it in our power to begin the world over again.

~Thomas Paine

———∞∞∞———

We can examine history in terms of the opposing forces of centralization and decentralization. When the printing press was introduced, Europe could have been viewed as a collection of frequently warring nations. However, its intellectual life was dominated by the authority of the Catholic church, which exerted strong control of education and disseminated its religious teachings to its congregations. The European nation states were kingdoms, often including assemblies of aristocrats which would later become parliaments.

The printing press decentralized the control of information and thus undermined all authorities. Just as the Protestant Reformation was a decentralization of the power of the Catholic Church, the American Revolution was a decentralization of the power of monarchy. The Revolution didn't merely eject the monarchy from the colonies, it undermined monarchies everywhere, principally because it offered the example of a king-free democracy that was also economically successful.

Other consequences took time. It wasn't until the second half of the 20th century that the European empires collapsed, ending finally with the collapse of the Soviet Union. On that particular battlefield, decentralization moved slowly, reduced to a snail's pace by the force of arms. There are other battlefields where the contest is still in progress: in the centralization of government and in the centralization of banking. We shall touch on those later.

In our view that was certainly the case for computer technology. It began life centralized, with the power of centralized technology dominant, beginning with the mainframe and continuing with the centralized power of juggernauts like Google and Amazon. Then came the blockchain, a technology that clearly pulls in the opposite direction.

161

The Silicon Revolution

Thirty years after the invention of the computer, during World War II, the MITS Altair 8800 appeared—the first personal computer available to consumers. Roll forward another 15 years to 1990 and you meet the first popular Windows PC. Seventeen years later, Steve Jobs launched the iPhone and completely disrupted the mobile phone industry. A few years later you could hold in your hand an iPhone containing enough computing power to happily run a laptop or desktop computer, or a mainframe from a decade before.

The magic of Moore's Law put once unimaginable computer power in the hands of over 2 billion people, almost a third of the world's population. You can add an estimated billion personally owned PCs to that. The applications populating these hardware devices distribute remarkable levels of personal capability. There have been wave after wave of them, incrementally increasing the power these devices deliver—making them an indispensable tool for normal life.

Silicon laid the foundation for a far more dramatic devolution of power to the individual than the printing press ever achieved. Silicon delivered the ability to record data, share data, process data, and own data to a massive population of enthusiastic users.

The Integration Revolution

In 1994, just when it seemed that the personal computing revolution was stalling, the Internet sprang up as if from nowhere. It changed the dynamic from decentralized personal applications to networked applications. The Internet spawned the dot-com boom, a wave of exuberant innovation that exploded from 1995 to the turn of the century, sending the stock market into the stratosphere. The NASDAQ composite index, dominated by dotcom and tech stocks, rose by over 500% in five years, only to collapse back to 1996 levels when the bubble burst.

There is a familiar pattern in such waves of innovation. They are driven by small business entrepreneurs, who vigorously pursue their dreams, test them in the market place and either prosper or vanish with their success or failure. The failures always outnumber the successes by a large margin and to the victors go the spoils. Looking back to the growth of the PC market, there were many small PC manufacturers, but only a few—Compaq and Dell being the main ones—made it big. The same happened with word processing, spreadsheet software, and presentation software. There were many early competitors, but in time Microsoft ate up the market. With the Internet there were many ISPs, many email businesses, many eretail sites, many auction sites and even quite a few search engines. But when the dust settled, there were far fewer.

An important aspect of the integration revolution was that it globalized computing, to the point where authoritarian regimes (China, Cuba, Iran and North Korea, for example) found it necessary to police and censor Internet usage, at least to some degree. Nevertheless most Internet users experienced freedom of information in a new way. Not only did a vast repository of information become available, it could be accessed in seconds.

Aside from the remarkable Wikipedia and Google Books (estimated at 25 million searchable books), there is now a vast array of free educational and news resources. Electronic publishing exceeded the possibilities of print publishing by many orders of magnitude.

Economies of Scale: The Big Battalions

The Vicomte de Turenne, a 17th century French military officer, noted that "God is on the side of the big battalions." Voltaire offered an alternative view, "God is not on the side of the big battalions, but on the side of those who shoot best." We can apply these two perspectives, both of which make an important point, to the commercial battles that arise in technology-based and Internet-based businesses:

– Amazon acquired its economies of scale very quickly, before it could be caught by anyone, first in the on-line book business and then across a much broader swathe of eretail and, again, with its investment in cloud computing.

– Neither Google nor Facebook had first mover advantage, but "they shot best," and acquired their economies of scale accordingly.

– Apple trailed in the dust behind Microsoft for over 30 years, but dashed past it, in terms of corporate value, in 2010. Steve Jobs turned out to be "a better shot" in the long run.

Economies of scale count and they count mightily. Currently they favor centralized business models and can be lethal to both large and small competitors. The businesses with the big battalions are under far greater threat from disruptive competitors than they are from direct competitors. Indeed they often don't notice disruptive competition until it is too late. If they notice it in time, they usually acquire the offending enterprise at a premium price. It's usually their best play; Facebook's acquisition of Instagram and Google's of Motorola Mobility serve as examples.

If we widen this perspective, we note that, as time has flown by, most sectors of the economy—banking, the auto industry, agriculture, retail, entertainment, and so on—have been consolidating and continue to do so. Where businesses are global, centralization is global. If we widen the perspective to include government, we see the same effect; government tends to get bigger, even when presided over by political parties that promised the opposite. And in most countries, government tends to act in favor of the commercial sector. In that area the rule of law is usually usurped by vested interests.

This is clearly the case in the US tech sector, where Alphabet (Google), Facebook, Amazon, Microsoft and Apple all spend substantial money on lobbying to influence legislation. However, their economies of scale depend far more on their ability to gather data and throw algorithms at it. They have exhibited considerable talent in doing so, although in many situations it is not their data.

The Blockchain Counter Revolution

In discussing the implications of the blockchain, we do not champion Bitcoin's implementation of it, just as in the discussing print technology, no-one thinks of Gutenberg's original print technology as the last word in printing. Bitcoin's blockchain technology has already been superseded by other approaches to the blockchain that achieve far lower transaction costs. There may even be other data structures, yet to be invented, that have the same virtues as the blockchain—bullet-proof wallet security, decentralized technology, decentralized consensus implementation, publicly viewable ledger, immutability, and smart contracts—but improve on it in some way. The revolutionary nature of the blockchain is in those those virtues, rather than their technological implementation.

As we have noted, blockchain technology can be applied in a surprising variety of ways. However, its versatility has been eclipsed by the speculative boom in cryptocurrency investment and the attendant price volatility. The wave of ICOs that washed over the crypto world in 2017 raised $5.6 billion, and despite the decline in the crypto market, that figure had more than doubled by June 2018.

Thus the blockchain disrupted the VC market through the ICO, a decentralized funding capability that could attract the participation of a legion of small investors. Despite the discouraging (and in our view sensible) intervention of the SEC in the US, the ICO is still a channel to funding elsewhere. Some US start-ups have taken their ICOs to friendlier climes. Particularly important in this trend will be the success of those projects that seek to decentralize the Internet, for they undermine national authorities like the SEC.

The Decentralization of the Internet

You could argue that the physical Internet is already decentralized, in the sense that it embodies a large number of independently owned components. But the Internet is also at the mercy of some critical centralized services: large communications infrastructure providers and ISPs, cloud computing resources, DNS services, email services and search. These services could be decentralized, and ultimately we expect them to be. The primary benefits would be twofold: first, decentralization would provide an Internet that is far more secure against hackers of every variety, and also more resilient to failure, and secondly, the governance of these services would be transparent and far less susceptible to exploitation.

Currently the most promising blockchain project in this area is the Inter Planetary File System (IPFS), a protocol and network designed and implemented from the get go as an open source project, to provide addressable peer-to-peer access to data. It is currently being developed in earnest by Filecoin. In short, IPFS could be used for web hosting, and it would defeat the Distributed Denial of Service (DDoS) attacks that are common and can even damage the DNS service on which the Internet depends. The DNS could be implemented on IPFS. A project called Nebulis intends to do just that. And these are not the only initiatives that aim to decentralize the Internet. The Safe Network, under construction by MaidSafe, has the same goals and, remarkably, was begun in 2006, before Bitcoin even existed.

The decentralization of data storage is an inherent part of these projects. Blockchain projects such as Storj and SIA specialize in this. From the users' perspective such an arrangement for data storage would be similar to Dropbox, resilient and reliable. The mostly likely difference would be that participation in the network would require you to provide your own computing resources as part of the arrangement.

Another blockchain-based project, Privatix, is dedicated to building a network of Virtual Private Networks (VPNs are secure peer-to-peer network connections) by enabling users to share their individual bandwidth. Such a network would eliminate both Internet censorship and threats of hacking and could conceivably cut VPN costs to ribbons.

A detail worth emphasizing here: decentralizing will happen through the sharing of personal or household computer resources. Since most of the expenditure on such resources (mobile devices and PCs) will be made anyway, a market for sharable resources is likely to rise up. Technology companies will see in this market an opportunity they cannot pass up. In fact, the precedent for this already exists: Bitmain, a 5-year old company that dominates the market for Bitcoin mining equipment. Its estimated worth is now $14 billion. From the individual's perspective, the logic of buying devices that major in and are designed for resource sharing is simple; they will pay for themselves reasonably quickly and can become a passive source of revenue.

Decentralizing Markets

A market is a centralized operation. Participants or their brokers come to a common hub to trade with each other. The market determines the current price of whatever is being traded, since previous trades are publicly displayed in real time. It might then seem that decentralizing such an operation would be difficult.

Nevertheless, LocalBitcoins, a business founded in Helsinki, Finland, to enable local street sales of Bitcoin, has been running successfully in a decentralized fashion since 2012. It has a web site on which those who wish to sell Bitcoin can advertise the prices they are willing to accept and buyers can then contact them and agree how to meet up and how to pay. Payment is usually in cash, or via on-line banking. LocalBitcoins also provides a reputation and feedback mechanism and even provides an escrow and conflict resolution service.

The idea of a trustless exchange protocol that could be used to establish a fully automated decentralized market for cryptocurrencies was first mapped out by Sergio Demian Lerner, although it was another year before what are called "atomic swaps" were invented. Simply explained, an atomic swap enables trading directly between two blockchains: between Bitcoin and Litecoin, for example. It requires the two parties to the trade to establish a shared "password" that only they know. This prevents a hacker from barging in and trying to steal either amount. The parties use the password to establish a private channel through which the wallet-to-wallet trade takes place. The interaction is governed by a smart contract and, when both participants have fully made the exchange, the channel is closed and the respective blockchains are updated.

The implementation of atomic swap capability began in September 2017 with Decred and Litecoin being the first two coins to support it. Since then at least 20 more have joined in.

There are significant advantages to using atomic swaps.

1. Atomic swaps are secure against any kind of hack, while crypto exchanges are obviously not, as shown by the record of Mt Gox, CoinCheck, etc.

2. Exchanges can have liquidity issues in respect of their stock of any given cryptocurrency, and this can delay a trade while the price moves.

3. As long as the blockchains involved in the swap have parties that wish to make such a trade, it will take place. (That may not always be the case.)

4. Centralized exchanges are registered in and governed by the laws of the country in which they are located. Atomic swaps are not subject to regulation.

Because atomic swaps are not governed by an exchange, there is no set price at which the trade must occur. Typically, it will be agreed between the parties with reference to the price on an agreed exchange, but that need not be the case.

The importance of atomic swaps is that they provide a template to enable the creation of other decentralized markets. With two cryptocurrencies it is relatively easy. From a software perspective it would be much more difficult to enable cryptocurrencies to be traded for stocks or other securities. It would be more difficult still to set up a retail operation that mirrored the Amazon market place or one that supported the kind of competitive auctions that occur on eBay. It would be difficult, but not impossible.

Decentralizing Bonds and Stocks

The US bond market is larger than the US stock market by about a third ($40 trillion compared to about $30 trillion). When the SEC began to fret about cryptocurrencies, it sought to distinguish between those that were investments (i.e. securities) and those that were not. As a rough rule of thumb, if the promotional material that accompanies a cryptocurrency ICO suggests that the token will increase in value, the SEC will most likely deem it to be a security, and require that it is only offered to accredited investors. An accredited investor is someone whose net worth (aside from the value of their primary residence) exceeds $1,000,000, or has an income above $200,000, or a joint income with spouse above $300,000.

A primary dynamic of ICOs is crowd-funding that anyone can participate in. However, the policy of the SEC has eliminated many people in the US, who might otherwise be interested, from participating in ICOs. The SEC is trying to protect the small investor from cryptocurrency scams, and as there have been quite a few, their policy has its positive side. Additionally it is worth noting that its only imposition on cryptocurrency exchanges is that they must follow Know Your Customer (KYC) guidelines. The goal is to discourage the use of cryptocurrencies in money laundering.

As a result of decentralization, we can expect bonds of every kind, corporate and government, to be issued as blockchain-based securities with an underlying cryptocurrency. The SEC regulations currently impede this development, but not in a way that prevents it. It prevents the crowd-funding of bonds, but not subsequent trading in a bond cryptocurrency once a bond issue has completed.

For those who are unfamiliar with bonds, here's a quick description: Bonds have an issue price, a defined interest rate which regularly pays interest amounts on the issue price, and a termination date. So if you buy a 5 year bond at a price of $1000, which pays out at 2.5% of the price annually, you receive $25 every year in interest, and your original $1000 is returned to you at the end of 5 years. (Some bonds are more complex, as the interest rate is linked to a variable interest rate index.)

If there were an underlying cryptocurrency linked to a specific bond, the bond could, for example, be priced at $1 and the value of the coin would simply fluctuate in harmony with the value of the associated bond. If you think about it, there is very little difference between a bond issue and an

ICO for a cryptocurrency, which is simply the issue of cryptocurrency. If the cryptocurrency is issued at a given price, pays a declared annual dividend at a particular interest rate and terminates at a given date, when all holders are reimbursed at the fiat price they originally paid–there's not much difference between that and a bond.

The rationale for crypto bonds is:

1. Bonds are well-defined. The investment proposition of a crypto bond is simple to understand, even to those who are unfamiliar with bonds.

2. Bonds could be crowd-funded. Because the crypto market is global, and small amounts could be invested, the market for the bond would be expanded.

3. Small transaction costs make crowd-funding feasible.

An important fact to note about bonds: if the issuing company goes broke, the bondholders have first call on its assets. They carry a risk, which is why the interest paid is higher for corporate bonds, but in the event of a failure, bondholders have priority over stockholders.

Most ICOs could be classified as stock IPOs—they are similar. They are carried out to fund a business, but are distinct from IPOs, in that the currency may have a designated function, aside from its investment potential. In our view, this is not a happy situation, because stocks are expected to be volatile, but currencies are not supposed to be. This distinction creates an unwelcome tension between the value of the currency as a speculative investment and its value in use. There is nothing to stop existing publicly quoted companies from issuing cryptocurrencies, but it is hard to think of a reason for them to do so.

More likely they will only get involved in blockchain activity by transferring their existing reward schemes (air miles, points or whatever) onto the blockchain. Consider, for example, air miles. How much is an air mile worth? With most carriers you can buy them directly, so you can make a guess, but estimates suggest that air miles costing 30 cents are generally worth no more than 10 to 15. And, as that estimate implies, there is no way to fix a specific price. If the carrier put their air miles on a blockchain, it would at least be possible to know their value, as all transactions would be transparent. However, the carriers probably do not want transparency.

As a natural consequence of blockchain technology, we can expect decentralized markets for stock to develop and the transaction cost of stock trading to reduce in a way that will challenge established brokers. We see no reason why you could not have atomic swaps of cryptocurrency for stock, as long as the stock you buy is registered somewhere in the cryptosphere, and its ownership instantly transferable. It is, after all just a market.

Incidentally, blockchain technology, used as a public ledger, would be an effective way to keep track of the derivatives market, which is larger than one's imagination can encompass, at an estimated $1.2 quadrillion. Given that the primary cause of the 2008 financial crisis was the proliferation of unregulated derivatives, it may make good sense to employ blockchain to track derivatives, in a controlled way. It could be used to monitor areas of potential risk (defaults, liquidity risk and the risk of "chain reactions" between holders).

The Banking Sector

In *Chapter 5, Let There Be Money*, we discussed, from the cryptocurrency perspective, three different kinds of money: *bearer money, account money* and *fractional reserve money*. We concluded that a cryptocurrency could replace *account money* because it fulfills exactly the same role as a checking account. We also concluded that cryptocurrencies could not replace either of the other two kinds of money. The broad conclusion then, is that national currencies will persist for the purpose of providing both *bearer money* and *fractional reserve money*.

Since the fundamental business of banking is to manage the market between savers and borrowers, it is likely that blockchain banks will be established, which simply record all their activity on the blockchain and manage every deposit or loan via a smart contract. This kind of transparency (and security) for banking would be welcome, at least to regulators.

The conundrum in bank lending is when to declare a delinquent loan to be a liability. Technically, all loans are performing assets until the borrower can no longer pay. Banks can fail because they "restructure" bad loans instead of calling them in, eventually reaching the point where their liabilities exceed their assets. A blockchain banking system could prevent that by implementing immutable smart contracts for all loans. It would diminish bank failures by reducing, or even eliminating reckless lending.

However, it would require the banking industry to take the blockchain more seriously than it has to date.

The major banks were quick to notice the emergence of cryptocurrencies and after a good deal of natural skepticism they acknowledged that the technology had its uses. With the sudden success of Ripple as an international payments mechanism, they realized that crypto was a threat. So now there are quite a few projects to create payment mechanisms, most notably IBM's initiative with Stellar Lumens. This is threatening news for the various financial businesses involved in making international payments and clearing checks. They will inevitably see their revenues shrink. But a far more fundamental application of blockchain technology will be needed to decentralize the banking industry away from its current centralized state. It will happen, but we expect it to take time.

Currently the banking regulation in the US is dominated by laws that are framed by lobbyists working on behalf of the banks. Until there is fundamental change, there is little possibility of the banking industry being decentralized in any way. The too-big-to-fail banks have no interest in downsizing, just as King George III had no interest in giving up his American colonies. As things stand, central banks are the "lenders of last resort" that preside over interest rates and act to stabilize the monetary system when the domain of fractional money encounters rough weather.

The viability of this arrangement will be called into question if a systemic banking failure occurs that the Federal Reserve cannot resolve by money printing. However, it will also be called into question if a reliable, non-inflationary cryptocurrency becomes popular to the point where millions of people begin to depend upon it, not just as a currency of account, but as a metric of value.

The Crossover Point

If that happens, businesses will want to set prices in that currency and many people will wish to be paid in that currency. The crossover point between fiat currency and cryptocurrency cannot occur until that happens. The only precedent we have for such a sea change is the period in the 19th and early 20th centuries, when the gold standard dominated the major currencies of the world and made monetary inflation impossible. The gold standard didn't fail, it was abandoned by governments who wanted to finance their wars. It will be far more difficult for them to move against a cryptocurrency which is entirely independent of their control.

If a fiat currency is thrust into competition with a stable and dependable cryptocurrency, the only way it will be able to compete is by offering a comparable level of stability, or by banning the use of the currency. However, prohibition may prove impossible. To understand why that may be, we need to consider the decentralization of government, which we expect to be a consequence of the second enlightenment.

The Drum Beat of the Second Revolution

The Enlightenment still casts its intellectual shadow over the modern era. Irrespective of political shenanigans, we still live in an age where rational science is the dominant source of knowledge. Economists may come and go, but Adam Smith's championing of free markets has not been dethroned by alternative thinking. Democratic government and the enfranchisement of citizens has increased dramatically in the past fifty years. The number of democratically run countries has doubled since 1970 and the majority of the world's population (over 4 billion) now live in democracies.

Despite the fact that the long march of democracy began with the first Enlightenment, it took until 1996 before the majority of the world's population was living under a democracy. The evidence suggests that economic success walks hand-in-hand with political liberation. In the 1990s, the economic failure of the Soviet Union, and the subsequent collapse of that sprawling empire, tipped the balance.

In our view the second enlightenment heralds a greater level of decentralization and a greater level of freedom than was achieved by the first. It has already begun among a few pioneers—cryptocurrency enthusiasts in the main—who have come to understand that winds of change are blowing across the world, creating individual freedoms that were previously impossible to achieve.

The two forces that will drive this are:

1. Self sovereignty
2. The immutable rule of law

The Decentralization of Identity: Self Sovereignty

In *Chapter 3, The Data Rights of Man*, we set out a Declaration of Data Rights. We discuss them here as a basis for defining self-sovereignty, the main point being that, as things stand, you do not necessarily own your identity. Your identity is conferred on you by the country in which you were born and supplemented by the digitally recorded events of your life—data over which you may or may not have control.

For the purposes of discussion, we classify data as falling roughly into three categories:

- **Credential Data.** These are data objects such as birth certificates, Social Security Numbers, driving licenses, and so on, which you own and which prove something about you, such as your age or that you are legally entitled to drive a car.

- **History Data and State Data.** These are data objects or data collections which record the activities and states of the individual data owner. It could be medical data, educational data, financial activity or even website activity. This is background data about you.

- **Title Data and Data Possessions.** Title data is data that proves ownership of a physical thing. A data possession exists entirely as data (like a cryptowallet or a photograph). These are possessions and thus their ownership may be transferable.

A primary goal of The Declaration of Data Rights is to draw a line in the sand, by explicitly declaring what an individual's data rights are in respect of the three types of data described above. What we provide here is merely a draft that we hope will be improved by others through criticism and intelligent suggestion. These are its first seven assertions:

1. The foundation of an individual's personal data is their identity. It is their personal property vouched for by a recognized validator. This fundamental credential data cannot be revoked.

2. Any additions or changes to this fundamental credential data are subject to verification by the original validator or its delegates. Such data may be subject to contract.

3. Secondary credential data may be provided by any validator who validates the individual's identity in the process of providing the credentials. Such data may be subject to contract.

4. The management of personal data by those incapable of managing it themselves is through a custodian or joint custodians. Incapacity can be due to age (as in the case of

minor children) and admits three possibilities: fully incapable, capable of managing data privacy and communications only, and fully data capable. Jointly owned data that is not subject to a formal contract is equally shared among its owners.

5. Explicit permission is required for the recording and retention of personal history data and personal state data, such as, for example, your current location. This may be subject to contract.

6. Personal history and state data is not transferable. It can only be rented by permission. This may be subject to contract.

In considering personal data, not only do we need to allow for the current situation of the individual in various countries in the world, we also need to think in terms of how it might change. Currently we live in a paper-based technology world. As a consequence, most personal credentials are documents, or digitally enhanced plastic cards of some kind. If, in the future, all credentials (birth certificate, wedding certificate, driving license, passport, debit cards, insurance certificates, club memberships, even the keys to your front door) are stored on a blockchain somewhere and are accessible electronically, none of those documents will be required.

At a traffic stop, for example, the police may wish to know only that you have a valid driving license and insurance, and are the owner of the car. If an exchange of data is legally required, given the context (speeding, for example), the exchange could be achieved automatically and be governed by smart contract. A similar process could occur when presenting an airline ticket and passport or other ID at an airport. The same model could apply to the presentation of credentials anywhere for any purpose, including going to the movies (digital movie ticket), paying for your groceries (digital debit card), or opening the front door of an office building (digital key).

The first three assertions in the declaration concern the importance of a validator. Currently, the primary validator in many situations is local or national government. The government provides validating documents for births, marriages, deaths, passports, driving licenses and so on. In a blockchain-based digital world, there is no necessity that such documents be provided by the government; the real necessity is that they be provided—immutably and securely—by a trusted, transparent, and auditable agency. Being a proven citizen entitles you to benefits of various kinds, which vary from country to country: education, health care, the right to be employed (and be taxed) and so on. The entitlement to such benefits is proven by credentials, which are usually acquired by reference to other trusted credentials. Thus, within a national context, there are validators that depend upon the trustworthiness of other validators, thus, forming a chain of validation. This arrangement could be replicated at every level by decentralized blockchain agencies which had the country's population as stakeholders and which funded themselves simply by charging for the service they provide. In other words, they could be taken away from government and decentralized.

The fourth assertion deals with custodianship. Clearly there is a need for this in respect of children, if for no other reason than that children will not appreciate the importance of personal data until later in life. However, the assertion also implies that custodians ought not to be intrusive, as children have a right to their data and privacy. Their data rights could and probably should be enshrined in a smart contract.

This assertion also mentions jointly owned data, which will be a common phenomenon. Wherever there is a transaction between two or more parties, the outcome is likely to create commonly owned data. If the ownership of the data is not confined to a single party, then its ownership should be shared.

Assertions 5 and 6 are self-explanatory.

Decentralizing the Rule of Law

The next four assertions in *The Declaration of Data Rights* concern data law.

7. Title data, in all its forms, is fully transferable from one owner to another.

8. Personal data is property, and it is subject to both property and copyright law.

9. Personal data has the right to be secure against unreasonable searches and seizures and it shall not be violated. No warrants shall be issued for access to it except upon probable cause, supported by oath or affirmation, and particularly describing the data to be accessed and searched, and the data to be seized.

This is the fourth amendment to the US constitution, repurposed to apply to data.

10. An individual's personal data cannot be used in evidence against him or her without the individual's permission.

    This is the fifth amendment to the US constitution, repurposed to apply to personal data.

The importance of the 7th assertion is that it directly (and in theory) transfers proof of ownership from the medium of paper documentation to the medium of digital documentation. In most circumstances, digital title will be far easier to exchange than paper title.

The subsequent three assertions are self-explanatory and difficult to oppose.

The Poster Child for Centralized Government

Authoritarian governments do not like digital freedoms. For decades now, the Chinese government has taken every measure it could think of to stifle political opposition by means of digital censorship. It has attempted to square the circle of achieving economic progress without enabling political progress. And it has done so by investing in technology. The blockchain is a definite threat to the Chinese government because of the freedoms it might unleash. Consequently, China is regulating the blockchain.

The Cyberspace Administration of China imposes the following regulative measures:

– Users of blockchain services must use their real names and record their national identification card numbers.

– Blockchain businesses must censor content, according to specific guidelines, and provide the authorities with access to user data.

China is also suppressing cryptocurrency trading and has the self-awarded right to review or simply take the data of any business. All of this violates the data rights we have set forth.

Broadly, China appears to have two distinct goals. On the economic front, it wants to discourage money from leaving the country and (rightly so) it sees cryptocurrencies as a mechanism for moving money offshore. On the individual level, China opposes data ownership and data privacy, and it opposes free speech—to the extent that the government heavily monitors everything that is posted to social networks. If you want self-sovereignty, move elsewhere.

None of this is particularly surprising, since China has been fretting about the political aspirations of its citizenry since the protests in Tian'anmen Square in 1989. With the birth of the Internet, China took the view that it had sovereignty over its domestic Internet, and it built "The Great Firewall of China," combining legislative actions with technologies to enforce that sovereignty. It limits or blocks access to a selection of foreign websites and limits the speed of cross-border Internet traffic. Foreign Internet companies that wish to operate in China have to abide by its regulations.

Depending on how it evolves, blockchain technology may prove very difficult for the Chinese regime to accommodate, because of its decentralized nature and China's entrenched centralization. Previously, technology was its ally, but this time it isn't. It is far easier, for example, to strangle the traffic on the current hierarchical Internet than it will be on a peer-to-peer Internet. In fact, it will probably be impossible for any kind of communications monitoring to identify and prevent VPN activity. But if China turns its back on blockchain technology, it will, we believe, suffer economically.

Decentralized Government

It has probably already occurred to you that democratic voting is destined for the blockchain; certainly the blockchain could implement a system that is secure, provides privacy, eliminates voter fraud and voter suppression, and replaces existing voting systems. A company called Voatz (Voatz.com) has built a system that is currently being piloted in West Virginia and has been tried out in 30 other small elections. The Voatz system uses provably secure mobile phones to record votes. Another player in this market is Agora (Agora.vote), which provides secure online voting to voters.

Issue-based voting is currently where there is the most activity. In 2018, in Japan, the city of Tsukuba introduced its own version of blockchain-based voting. Voters used social security identifiers, voting not for individuals, but for or against various local development programs. In Moscow, a blockchain-based app called Active Citizen has been used to cast votes on local municipal decisions. Liquid Democracy proposes a scheme that will make it possible for voters to wield more influence within the US system, by voting directly on an issue, or by delegation. If it sounds like these are the early days of blockchain democracy, it is because they are.

Nevertheless, from a technology perspective, the nation state is ceasing to have the relevance that it once had. The blockchain is subverting it in favor of a globalized, borderless, and increasingly decentralized ecosystem. The globalization ushered in by the Internet delivered the bulk of its rewards to large corporations and energetic start-ups that became large corporations. The blockchain is not their friend. Instead, the blockchain brings globalization to the individual and to small groups of collaborators.

Whatever its flavor, government is a monopoly of a kind. Those who curry its favor cozy up to it in whatever way the system allows, whether through lobbying, trading favors, or direct bribery. When you investigate who is doing what and how, you soon discover that many of the back-door relationships are between government and would be commercial monopolists. Despite the political championing of free market competition, large corporations know that genuine competition is a commercial threat, so they do whatever they can to stifle it, employing all the anti-competitive tactics available to them. Their shareholders would never forgive them if they didn't.

The blockchain revolution challenges this established order, and, with the alternative of a decentralized economic order, shakes its foundation. It levels the consumer v. provider playing field by providing an alternative economic arrangement, where consumers have the power to become stakeholders in enterprises that are governed by transparent smart contracts and their blockchain votes. The watchwords are no longer "liberty, equality and fraternity;" they are "security, transparency and trust."

Towards A Decentralized World

There are many questions to which we cannot yet know the answer.

– Which of the crypto currencies will dominate? It's too early to say.

– Which of the current crypto currency businesses will prosper? Too early to say.

– How will the blockchain reform government? Too early to say.

– Will the blockchain challenge authoritarian governments? Too early to say.

At the time of writing, all that the blockchain has proved is that it could provoke an investment mania, excite a whole new generation of technological innovators, and give birth to thousands of new businesses across the world. It is not without its consequences, but neither is there any clear indication of how this revolution will unfold.

A number of trends need to bear fruit before we can achieve that clarity. However, we know what those trends are, and we have described them throughout this book. We summarize them now, in the hope that they will stick in your mind as the nascent world of the blockchain gradually impinges on ours:

- Blockchain technology will be adopted internally or between businesses fairly quickly. The utility of a shared ledger system will become apparent and it will be particularly common in supply chain applications. However, this is likely to have little direct impact beyond the business-to-business world.

- The blockchain will replace the current infrastructure of the Internet, replacing it with a peer-to-peer organization. It may take several years just to establish a foundation, but it will happen. And when it does, the blockchain will father a new form of cloud computing, where individuals participate with their personal and home devices.

- Many blockchain business ideas depend upon individuals owning and managing their own data. The existing data giants will not be humbled until that happens. It is unlikely to be a fast adoption process, simply because most Internet-connected people do not yet understand the importance of data ownership. Most likely this will be driven by the emergence of a "killer application" for data ownership.

- The use of cryptocurrency for money transfers has already begun, but is still a niche application. As the cost of exchanging one cryptocurrency for another falls, the large banks will discover that they have an unwelcome competitor. Nevertheless, it will have only a minimal impact on their revenues.

- The use of cryptocurrency wallets as a replacement for checking accounts will not occur quickly and is most likely to happen, initially at least, through a fiat-linked stable currency. The impediment to this is the requirement for the cryptocurrency to be a metric of value, and this will happen first with cryptocurrencies that are linked to fiat currencies.

– The banking sector will ultimately be severely disrupted by the uptake of cryptocurrencies, but the process will take time. It requires large amounts of fiat currency to be swapped for cryptocurrency. When this starts to occur, the trend will become self-reinforcing. The value of the fiat currency will fall as the value of the cryptocurrency rises.

The political influence of the blockchain will be the last to make its presence felt, largely because the political tide requires a population of users (colonists if you like) who are already enjoying the freedoms that the blockchain provides and who collectively rebel against the aging economic regime that seeks to rule them.

The blockchain is the technological foundation of a political force, which can, and in all probability will, undermine the current political world order and replace it with a different one.

A Recapitulation

The major arc of this book is the clear parallel between the extensive changes wrought by the printing press—which we can think of as the distribution of the power to publish—and the extensive changes that are being wrought by the distribution of computer power, which recently gave birth to the blockchain and its associated technologies. The hypothesis is that these two distinct revolutions are following the same pattern.

The first led to dramatic changes in language, in religion, the intellectual authority of the Catholic Church, a growth in literacy, an explosion of publishing, an explosion of scientific activity and thinking, economic revolution and political revolution—leading to the founding of America.

Where the second will lead is still uncertain. Up to now it has led to increasing centralization, particularly in respect of aggregations of computer power, data storage and the analysis of data. In recent decades, increasing amounts of personal data have become digitized and stored online. What has gradually evolved is a data economy, where the largest and most valuable companies are those that have best managed to apply analytical techniques to very large accumulations of data—personal data.

The Counterforce

A counterforce that opposes the exploitation of personal data is gradually building. An aspect of this is EU General Data Protection Regulation (GDPR), which confers data rights on European citizens and applies world-wide. In our view these regulations can be thought of as a "good start," but are insufficient. We argue that data ownership is a fundamental human right and we propose a set of formal data rights that could and should be adopted.

We note that it is technically possible, in most circumstances, to build applications that allow data to be used anonymously. Strange though it may seem, businesses do not need to know an individual's personal identification details, they only need to know how to make contact. Blockchain technology is capable of implementing such applications and preventing the exploitation of personal data.

We identify three distinct types of personal data They are:

- **Personal Credential Data:** data that is used to validate your identity and entitlements.

- **Personal History Data and State Data:** data about you that you own and over which you should never be able to lose control.

- **Personal Title Data and Data Possessions:** tradable data whose ownership can be transferred.

Money

We examined the history of money from prehistoric times forward. The conclusions we reached are:

- Historically, coinage (however imperfect) has at times been a store of value, in that a gold coin is gold, which has intrinsic value. There have been other forms of money that have intrinsic value, such as cigarettes in post WWII Germany, or grain, or even livestock.

- Coinage and paper money are challenged (from the outside) by fraudulent actors but also (from the inside) by the "corrupt" governance exerted by governments. The gold standard did not fail, it was abandoned by governments who found the discipline it imposed intolerable.

- The final country to abandon the gold standard was the US in 1971, when President Nixon cut the link between gold and the dollar. Now no national currencies are backed by the intrinsic value that gold represents.

- The major problem of national currencies is governance. In general, governments cannot be trusted not to debase their currencies. History demonstrates that eventually it happens.

Analytically, we can think in terms of there being three types of money at a national level. Note that this categorization is new, deduced from considering the possibilities of a cryptocurrency. The three categories of money are:

- **Bearer money.** This is money that has no knowledge of who owns it. Whoever carries it is considered its owner.

- **Account money.** This is a store of money held somewhere on the owner's behalf.

- **Fractional reserve money.** We can think of as "invented" money, the product of fractional reserve banking. Where, for example, a bank takes a deposit of $1 million and lends out $10 million, using just the $1 million as a reserve against default.

How money supply is measured varies between nations, but in general the measures: M0, M1, M2, and M3 are used. These measures do not distinguish between the types of money that we have described.

Cryptocurrencies

Cryptocurrencies are **account money**. A cryptocurrency wallet is equivalent to a personally controlled bank account. By their nature, cryptocurrencies cannot be **bearer money** or **fractional reserve money**. The cryptocurrency enthusiasts who believe that cryptocurrencies will replace fiat currencies are dreaming. They will be disappointed.

The seven characteristics of a currency are as follows:

- It is a medium of exchange.
- It must be portable.
- It must be relatively inexpensive to create, maintain and use.
- It must not be easy to forge or steal.
- It must be easily divisible.
- It must act as a metric of value.
- It requires governance.

The advantages fiat currency has over cryptocurrency are as follows:

- Fiat currencies are superior mediums of exchange, because more people readily accept them and they are easier to use.
- Fiat currencies are the de facto metric of value. Most cryptos are too volatile to be a good metric of value.
- Fiat currencies are slightly more portable, but crypto can be almost as portable via card and phone capabilities.
- Fiat currencies can be **bearer money** and also **fractional reserve money** as well as **account money**; crypto can only be **account money**.

The advantages that cryptocurrency has over fiat currencies are as follows:

- Crypto can and generally does has far lower transaction costs than fiat.

- Crypto provides far greater divisibility than fiat.

- Crypto appears to be immune to fraud.

- Crypto is international, whereas fiat is not.

- Crypto governance is far superior to fiat governance.

In our view, the only possibility for a cryptocurrency to displace a fiat currency in any practical way would be if the cryptocurrency linked its value directly to a fiat currency. Even then it would, at a maximum, only displace the *account money* within an economy.

For example, *account money* makes up roughly 14% of the US money supply. Insofar as any cryptocurrency could replace the dollar, it could only replace 14% of the supply of dollars.

Blockchain Technology

We described and reviewed blockchain technology. Aside from the blockchain itself we drew attention to:

- **Smart contracts.** These are stored on the blockchain, and can be thought of as application software that runs on the blockchain.

- **Zero-knowledge proofs.** A zero-knowledge proof is one where a person can prove to another that they have specific knowledge without revealing the knowledge. Such proofs have multiple applications, as they enable anonymization and can be used to check credentials without the credentials needing to be shown.

- **The Interplanetary File System.** This offers an alternative file system to the one defined by the HTTP protocol and has many advantages.

- **Solid.** The Solid project, initiated by Prof. Tim Berners Lee, is an attempt to decentralize the web, but do so in a way that is compatible with the web as it exists today.

When considering the potential of blockchain technology, all of these aspects of blockchain technology need to be taken into account.

Blockchain Applications

We examined categories of blockchain applications that have already proved themselves to some degree. The following is a list of those covered:

- New currencies focussed on inexpensive payment transactions and new currencies focussed on micropayments.
- Cryptocurrency banking.
- Exchanges, ICOs, crowdfunding.
- Businesses (of all kinds) using tokens as a reward or payment capability.
- Record keeping.
- Video gaming is a natural application for tokens and likely to see phenomenal growth.
- Consumer friendly advertising.

Blockchain based advertising models are of particular interest because they constitute an example of individuals being able to monetize their personal data. It is likely that, as individuals begin to appreciate that they have data rights and begin to enforce them, a multitude of opportunities to trade (rent) personal data in a controlled way will develop.

We estimate that (currently) the personal data of a US resident is worth, on average, somewhere in the region of $2000—$3000 per year. This figure considers only the following means of estimating personal data value: Its value to a digital advertiser for ad-targeting, it value to a data thief, its value in calculating credit scores and the value of an individual's data tracks.

We note, however, that currently there are no mechanisms whereby individuals can collaborate to better monetize their data by, in particular, using the kind of data mining algorithms that are used against them.

The Cryptocurrency Treasury Function

We examined the mechanisms available to a cryptocurrency business to stabilize its token value. In our view an unstable token value (one that fluctuates wildly) is an impediment to a cryptocurrency business. We concluded that most cryptocurrency businesses need to create a treasury function.

The primary token management goals of a crypto business will be to keep the token value stable, or to have a slowly rising token price. Most crypto businesses can know (or estimate) the value of their token far more accurately than any other agency and, in particular, far more accurately than a cryptocurrency exchange. They can estimate the token's value in respect of its practical use.

Consequently, a crypto-business could stabilize its token by publishing its estimated value (in realtime) and automatically and transparently buying or selling when the value fell unreasonably low or rose unreasonably high. We described the specific tactics that a crypto-business could pursue to impact the value of its token as part of the treasury function.

Our Possible Futures

At the time of writing, we are in a lull in respect of the popularity of cryptocurrencies and the enthusiasm for them among the general public. There are some developments that we can expect to occur:

- Blockchain technology will be adopted and used, where appropriate, by mainstream businesses, particularly in supply chain applications. It will supersede the current infrastructure of the Internet, replacing it with a peer-to-peer organization. The blockchain will father a new form of cloud computing, where individuals participate with their personal and home devices. This will not happen quickly, but it will happen.

- As is already apparent, there will be many innovative businesses that base themselves on the blockchain.

- The use of cryptocurrency for money transfer has already begun, and the large banks have already discovered that they have an unwelcome competitor. Nevertheless, it will have only a minimal impact on their revenues until a popular stable cryptocurrency emerges. The use of cryptocurrency wallets will then begin to replace checking accounts. Ultimately, the banking sector will be severely disrupted.

- Individuals will need to struggle to establish the ownership and management of their personal data. Most people do not yet understand the importance of data ownership. Until they do, they will be the victims of data aggregators, who profit from personal data exploitation. They may also be the victims of centralized government initiatives, such as those pursued by China and other totalitarian regimes which seek to prevent personal data ownership.

A Recapitulation

The Data Rights of Man, and their Political Implications

If the analogy we drew between the printing press and personal computer power is accurate, the political implications of the blockchain will make their presence felt. We believe that this has begun to occur and, in the spirit of championing that dynamic, we have invoked the ghost of Tom Paine. And, for good measure, we have formulated a set of Data Rights, inspired to some degree by the US Constitution.

If history is anything to go by, it will take time for the revolution we envisage to erupt. To have force it will require a significant population of users (colonists if you like) who already enjoy the self-sovereignty that the blockchain enables. They will be opposed everywhere by the current aging economic regime and the plutocracy whose wealth will be diminished if such freedoms spread.

There will be a popular revolution. And if history rhymes, then that uprising will bear some similarity to the one that began in 1765. This time, however, instead of undermining the monarchs and emperors of the world, it will undermine the very idea of a nation state.

And, it will be replaced with something distinctly different.

Acknowledgements

The author acknowledges the contributions made to the contents of this book by suggestions from and discussions with the following people: Charles Silver, Raj Lakhani, Steven Wilkinson, David N ayer, Michelle Ray, Shane McDougal, Tim Negris, Paula Schmidt, Judy Ryser and Fox RedSky. Wikipedia deserves acknowledgement for the contribution it made—it is a remarkable, and typically, an excellent research resource.

The author also acknowledges the editing work performed by Judy Ryser, Paula Schmidt and Jude Bloor. Without their work, the text would have been far more troublesome for the reader.

Author's Biographical Notes

Robin Bloor was born in 1951 in Liverpool, UK. He obtained a BSc in Mathematics at Nottingham University and took up a career in the computer industry, initially as a software developer, later a project manager, and then a database consultant. From 1989 onwards, he became a technology analyst. He has been an author ever since.

In 1999, he authored a book entitled The Electronic B@zaar, which provided a guide to the dotcom revolution that was unfolding at the time and became a business best seller in the UK. In subsequent years he co-authored three For Dummies books, covering different technological topics.

In 2002 he was awarded an honorary Ph.D. in Computer Science by Wolverhampton University in the UK.

In 2015 he co-authored The Algebra of Data with Professor Gary Sherman, a book that introduced a wholly new mathematical approach to data. He has also published several niche books on the writings of G I Gurdjieff.

He is a frequent blogger on technology matters (on Medium) and has focused his attention on the cryptocurrency world in recent years. He is also associated with several cryptocurrency businesses. In particular, he is an Advisory Board Member for Permission.io and is involved with another cryptocurrency project that is currently in stealth.